CBT

Cognitive Behavioural Therapy

CBT

Cognitive Behavioural Therapy

YOUR TOOLKIT TO
MODIFY MOOD, OVERCOME
OBSTRUCTIONS AND
IMPROVE YOUR LIFE

**Elaine Iljon Foreman and
Clair Pollard**

ICON

Published in the UK and USA in 2025 by
Icon Books Ltd, Omnibus Business Centre,
39–41 North Road, London N7 9DP
email: info@iconbooks.com
www.iconbooks.com

First published in the UK in 2011 by Icon Books Ltd

ISBN: 978-183773-267-8
eBook: 978-183773-268-5

Typeset in Adobe Caslon by SJmagic DESIGN SERVICES, India.

Printed and bound in the UK.

About the authors

Elaine Iljon Foreman is a Chartered Clinical Psychologist specializing in the treatment of anxiety-related problems. Her clinical research into Cognitive Behavioural Therapy techniques, developed over 30 years, has meant she is regularly in demand by the media for her expert contribution. Elaine's research into the treatment of anxiety and particularly fear of flying has generated invitations to present her research in this field across Europe, the Americas, Australia, the Middle and Far East.

Dr Clair Pollard is a Chartered Clinical Psychologist and is accredited by the BABCP (British Association for Behavioural and Cognitive Psychotherapies). Clair works in the NHS with adults with mental health problems and for a charity called The Back-Up Trust, working with people with spinal cord injury. Clair has a particular interest in issues around adjustment to disability and in post-traumatic coping and growth.

Acknowledgements

A book's value is based on its usefulness to the reader. Ian McLeod, Visiting Professor of Law at Teesside University has provided invaluable input to enhance clarity and simplicity, contributing greatly in the usefulness stakes. Our very grateful thanks are expressed to Ian and also to Duncan Heath, our editor, and his team at Icon Books for their assistance in making this book amongst the most useful in enabling people to change.

We would also like to acknowledge the following people who have made particularly significant contributions to the development and practice of CBT and on whose work we have drawn:

Aaron T. Beck, Judith Beck, Gillian Butler, David Clark, Albert Ellis, Paul Gilbert, Ann Hackman, John Kabat-Zinn, Paul Kennedy, Marsha Linehan, Stirling Moorey, Christine Padesky, Paul Salkovskis, Maggie Stanton, Richard Stott, John Teasdale, Adrian Wells and Mark Williams.

Dedication

No man is an island, it is said – nor woman either. We would like to thank the rocks surrounding us (you know who you are) who have given such support and encouragement in our writing – and in everything else. On your solid foundation we can stand firm and strong, deal with life, and enjoy!

Authors' note

It's important to note that there is much frequently-used research employed in cognitive behavioural therapy. Where we know the source we have been sure to reference it, but our apologies here to the originators of any material if we have overlooked them.

Contents

CHAPTER 1

Introducing CBT

*Men are disturbed not by things but by the views
which they take of them ... when, therefore, we are
hindered, or disturbed, or grieved, let us never blame
anyone but ourselves: that is, our own judgments.*

Epictetus, Greek philosopher

COGNITIVE BEHAVIOURAL THERAPY: WHERE DID IT COME FROM?

Some readers may be familiar with the name of Pavlov, and his early experiments in the 1900s looking at the way in which dogs can be 'conditioned' to salivate at the sound of a bell. What many people don't know is that Pavlov was actually studying the digestive system of dogs and just happened to observe this 'conditioned reflex'. However, he opened up a whole new field of study, allowing new insights into understanding the way in which animals learn. From there, it was but a small step to transfer this knowledge from four-legged to two-legged creatures. So the field of **behaviour therapy** was born. It stemmed from applying the principles of learning theory to shaping the behaviour of first animals and eventually humans, looking at ways in which altering behaviour might help alleviate psychological 'disorders'.

Dr Aaron Beck, originally a psychoanalyst, is generally credited with founding **cognitive therapy** in the 1970s. Working with depressed patients, he noticed they experienced a series of spontaneous negative thoughts, which he called **automatic thoughts**. He divided them into 3 categories: negative thoughts relating to the self, to the world and to the future. Working to identify and challenge these thoughts enabled patients to re-evaluate them more realistically. The result? The patients felt better and showed positive changes in behaviour. They became able to think in a more balanced, realistic way; to feel better emotionally and to behave in a more functional manner. The key concept of cognitive therapy concentrates on how we process information, organize it, store it, and relate new information to old. In cognitive therapy we focus on understanding the way in which humans think and applying these principles to the treatment of psychological disorders.

In the 1970s and 1980s furious debate raged over whether behaviour therapy or cognitive therapy held the key to understanding and overcoming psychological difficulties. Eventually, although there is no general agreement of exactly when, it became clear that this wasn't a contest with a winner and a loser. People neither operate on purely behavioural principles, nor live their lives based purely on thinking. It is in this realization that we find the roots of CBT.

There is an apocryphal tale of a paper written in the early 1990s. It was about understanding the nature and treatment of a particular anxiety disorder. The author believed in its importance (which author doesn't?!) and was convinced that the research had international significance. But there was a conundrum – two world conferences appeared to address the subject matter. One was the World Congress of Behaviour Therapy, the other the World Congress of Cognitive Therapy. The two were apparently miles apart, the conceptual separation reflected geographically, with one taking place in Canada, the other in Australia!

How to make a decision? As a behavioural experiment, the author sent an identical abstract to both. The result? BOTH were delighted to accept the paper as truly representative of research in that field! This is actually no shaggy dog's tale – the author in question is the first author of this book!

So what happened the following year? A historical first took place – the first World Congress of Behavioural and Cognitive Therapies. This fortunately meant that from then on, Elaine only had to submit her research to one conference, instead of two!

Modern 'Cognitive Behavioural Therapy' – the CBT which this book explores – applies principles of both schools of thought to the treatment of psychological distress. It looks at the way in which our thoughts, emotions, behaviours and physical states all interact to cause and maintain difficulties. Since we know all of these factors interact, it follows that altering any one of them will have an effect on the others. CBT focuses on the way we can change patterns of thinking and behaviour in order to feel better.

Now, nearly 20 years later, we are moving into what is called the 'third wave' of CBT. Instead of just 'mind' and 'behaviour', CBT is moving into domains previously addressed primarily by other traditions, in the hopes of improving both understanding and outcomes. Thus third-wave therapies include concepts such as mindfulness meditation, acceptance, values and relationships. The emphasis in these newer ways of working is less on changing the content of thoughts; rather it is on changing our awareness *of* and relationship *to* thoughts. If you are interested in reading more about this see our list of suggested resources in Chapter 8. So, where are we now?

WHAT IS COGNITIVE BEHAVIOURAL THERAPY?

Cognitive Behavioural Therapy (CBT) techniques have been developed from extensive research. Studies indicate that treatments for psychological disorders based on CBT principles:

- Are as effective as medication in treating many psychological disorders and often *more* effective in effecting long-lasting change and preventing relapse

- Are particularly effective for common mental health problems such as anxiety, depression, panic disorder, phobias (including agoraphobia and social phobia), stress, eating disorders, obsessive-compulsive disorder, post-traumatic stress disorder and difficulties with anger

- Can help if you have a low opinion of yourself, or physical health problems like pain or fatigue

- Can also be useful in helping to manage more severe mental health problems like bipolar disorder (previously called 'manic-depression') and psychosis.

CBT works on the principle that our behaviour and emotions depend to a large degree on our perception of what we understand is happening. What we think and anticipate can greatly affect our reaction to events and people. Having understood what you are thinking and how to deal with your thoughts, it is possible to train yourself to respond in a different way. This new style of thinking and behaving can then lead to a potentially more satisfying way of life, becoming part of your normal lifestyle.

CBT uses practical techniques and exercises to help you make lasting changes in the way in which you think and behave in order to help you feel better.

REMEMBER

In this book we concentrate on the practical skills and techniques that have been developed using the principles of CBT. They are presented in a format that is easy to use, so you can develop the life and lifestyle that is right for you. We have focused on the more common areas of mental distress and discomfort experienced by a large number of people.

Please note: If at any point when you are working through this book you find that things are getting worse rather than better, do seek professional help immediately. Likewise if your mood drops and you start to feel overwhelmed, ensure that you see either your GP or one of the mental health professionals described in Chapter 10.

CHAPTER 2

The ABC of CBT

REMEMBER: THE ABC OF CBT

Antecedents – Beliefs – Consequences

A = the **A**ntecedent, trigger event or occurrence which appears to lead to an emotional reaction.

B = our **B**eliefs, thoughts, interpretation or evaluation of that event and its possible causes or meaning.

C = the **C**onsequences of that way of viewing the event – our emotional or behavioural reaction to it.

Imagine you are in bed at night, alone in the house and you hear a sudden noise downstairs. This is the antecedent, the triggering event – an **A**.

You might think you'd know immediately how you'd react or feel in this situation. Actually, our feelings and reactions depend entirely on how we interpret the **A**. Look at the three possibilities below:

1. You might think: 'Oh gosh there have been several burglaries in this area recently, I bet it's them'. This would be a belief, or thought – a **B**. It might follow that you'd feel scared or even angry. This would be a consequence, or reaction – a **C**. Your behavioural reaction (another **C**) might be to hide under the covers or to call the police.

2. You might think: 'That's my son coming home late again and crashing around waking me up – third time this week – he's always so thoughtless!' – a very different **B**. In this case your reaction (**C**) might also be quite different. Now you might feel very angry and frustrated and your behavioural **C** might be to shout at him or try to impose some kind of sanction.

3. But you might think: 'Ahh! That's my lovely partner returning earlier than expected to surprise me because I was feeling a bit low today. How very sweet!' Then your feelings (**C**) might be loving and positive and your behavioural reactions (more **C**s) equally so!

So, in each of these scenarios the **A** is exactly the same. The **C**s are all completely different. What makes the difference? The **B**s, our beliefs! The way in which we think about the situation determines the way we feel about it and react to it.

Of course in real life things are more complicated. Our beliefs are influenced by myriad factors including our upbringing, education and past experiences. Behavioural and emotional **C**s in one situation feed into the As and Bs of other situations, and so on. However, bearing in mind these simple principles can help us understand and then make changes in many areas of difficulty.

Throughout this book we'll be illustrating how it's our Bs (beliefs) that largely cause the stressful Cs (consequences), not necessarily the actual situation itself. So, if someone isn't stressed about meeting important deadlines, giving a presentation or meeting new people it's because they *believe* they'll cope well and therefore don't predict any awful consequences. The fact that they're not stressed in this way can then become a self-fulfilling prophecy, in that it will cause them to behave and react in positive ways which might actually make a successful outcome more likely. When we hold overly negative beliefs the opposite can happen.

A key part in the process of challenging negative beliefs is to question the commands which say you *must*, *ought*, *should*, or even *have to* achieve a particular outcome. Where do these commands come from? Do they just pop automatically into your head, or are others telling you these things? If they're from others, is there a reason you have to agree? Are others necessarily infallible? What would happen if you did fail? Would it really be that unbearably awful? Could you be exaggerating the outcome? And if it did happen, is there a way you *could* bear it, even if you didn't like it? After all, there's no law which says you *have* to like it! Challenge those previously unquestioned assumptions. Ask yourself how one missed deadline means you are a complete waste of space at work. Isn't that a bit unfair on yourself? Would you judge others like this?

In becoming more aware of the beliefs that are driving your reactions and behaviours you can then make those beliefs more balanced, realistic and flexible, less demanding and no longer so absolute. When beliefs are modified you usually find that you feel emotionally and physically different. This actually enables you to evict catastrophizing and its companion, procrastination, and get on with the task in hand. The consequence? You normally find yourself feeling ever so much better than before.

Have another look at the quotation from Epictetus at the start of the introduction. Amazing how it's absolutely spot on, encapsulating the latest developments in CBT, despite being said over 2,000 years ago.

But identifying the **A**ntecedents, and the **B**eliefs, and then actually challenging them, both in your mind and your behaviour, like so many things, is much easier said than done.

It's very important to tell yourself that like any new skill, learning your ABC takes a while. It will take time before you know it by rote, and can incorporate it automatically into your daily routine.

USEFUL TIP
Think of this book as your mental workout – after all, you are 'working it out', aren't you? You really can sharpen, tone up and keep your mind fit by regular workouts at the mental gym. When you go to this space, your private mental gym, that's when you practice challenging unhelpful, mood-disturbing, distorted thinking. Your workout strengthens you, as you develop new qualities and performance-enhancing, stress-reducing, life-improving beliefs.

Managing anxiety

From ghoulies and ghosties
And long-leggedy beasties
And things that go bump in the night,
Good Lord, deliver us!

Traditional Cornish prayer

UNDERSTANDING ANXIETY

The internet offers around 46.5 million answers to what anxiety is all about! So that you don't have to go through them all (at 5 minutes per website that's around 450 years of your time) in this section we condense it down to the basic essentials:

- What it is
- Where it comes from
- What forms it takes and, most important of all …
- What you can DO about it.

WHAT IS IT?

Anxiety is often described as a feeling of worry, fear or trepidation. But it's much more than just a feeling. It encompasses **feelings or emotions, thoughts and bodily sensations**.

try it now

You might be more sensitive to one or two of these. Remember when you last felt really scared? Write down what you remember noticing, and then look at the examples we have given. Don't worry if one column's blank – it's common not to notice everything when you first start looking at your emotions, thoughts and physical feelings.

Situation when you last remember feeling terrified		
Physical sensations What happened in your body?	Emotions What did you feel?	Thoughts What went through your mind? Words? Pictures?
Examples of typical reactions		
Heart racing, sweating	Feeling absolutely petrified	What will happen next? Will I have a heart attack? Will I look like an idiot?

Occasional anxiety is absolutely normal within our everyday experience. If you didn't feel anxious, ever, *that* would be something to worry about! Life presents us with challenges, which we aren't always confident we can handle, so a degree of anxiety is natural. The challenges can be stressful events including actual danger, happening in the real world, and/or the things our minds conjure up, such as what *if* a catastrophe *did* occur – like meeting those ghoulies and ghosties which we mentioned at the start of this chapter.

FEELINGS OR EMOTIONS

When we experience severe anxiety we usually feel terrified. While sometimes it is quite straightforward to identify what it is that we are scared of, at other times we just get an overwhelming feeling of panic. But whether you love or hate this feeling depends to a great extent on your personality and the context.

Believe it or not some people seek strong sensations, and for these people sometimes the more powerful, the better! Experiencing high anxiety can be pleasurable, even though that might sound peculiar. Think of horror films, amusement parks or 'extreme sports' holidays. Certain people love the adrenaline rush these activities provide. The key is that usually the enjoyment is linked to it being a time, place and activity that they have chosen. They would probably be less enthusiastic about something that was happening to them uninvited, unwanted, out of their control and downright dangerous!

THOUGHTS

We all usually try to make sense of our environment, and to understand what is happening to us. It can be really frightening not to know what is happening, and to anticipate that whatever is going to happen next will be even worse. Anyone experiencing feelings of panic and terror is likely to try to figure out why it's happening, and what it means. How we make sense of our world is what tells us whether it is safe or dangerous. Shakespeare neatly summed this up, writing in *Hamlet*, 'there is nothing either good or bad, but thinking makes it so'.

So the link between thoughts and emotions is already becoming apparent – if you *think* something is *really* dangerous, you are likely to be seriously scared of it. People watching a horror movie are less likely to enjoy it if they then start looking out for aliens and monsters when they leave the cinema, while those who recognize it as being 'only make believe' can safely enjoy the scariness in the confines of the cinema, knowing that in reality there are no such dangers.

BODILY SENSATIONS

It can be quite astonishing to discover how many different sensations can be triggered by anxiety and how many different parts of the body can be affected. You may get just a few of these or most of them. The most common sensations are:

- Your heart may beat faster and harder

- Your chest may feel tight or painful

- You may sweat profusely

- You may tremble or have shaking arms and legs

- You may have icy cold feet and hands

- You may have a dry mouth

- You may have blurred vision

- You may need to go to the toilet or have a churning or fluttering stomach

- You may have a horrible headache

- You may feel that you're 'not really here' or that you are somehow out of your body, looking down on everything, detached from your surroundings

- You may feel as if everything is very unreal

- You may feel dizzy, light-headed or faint

- You may feel you have a lump in your throat or that you can't swallow

- You may feel nauseous – you may even vomit

- You may feel tense, restless or unable to relax

- You may have general aches and pains.

As we mentioned, it is normal to experience anxiety when we feel we are in danger. Your body responds with the 'triple F' reaction, Fight, Flight or Freeze. It's a really important automatic response – your body does it all by itself. The 3 Fs are linked to the survival of our species over the years. Take the example of disturbing a hungry wild animal out in the bush. Depending on both you and the type of animal, you might try to fight it, to run away as fast as you could, or to keep stock still in the hope that it had poor eyesight and wouldn't charge at you. Which of the 3 Fs do you reckon you'd choose?

In situations you perceive as dangerous, your body produces a whole range of chemicals (including adrenaline) which trigger all of the physical symptoms above. These bodily changes are what have helped the human race to survive. The chemicals released cause physical changes which enable us to run far faster than otherwise, have greater strength, and generally have a better chance of defending ourselves and our loved ones. That's great for an objective danger like a wild animal, but not particularly helpful when the perceived danger is more of a social one, like being afraid you will make a fool of yourself or a (most likely unfounded) fear of a physical catastrophe such as having a heart attack or brain haemorrhage.

In a moment we will go on to look at different specific types of anxiety problem. Each links to a range of thoughts about what is happening. So for instance, if you suffer from panic attacks you'll probably fear that when you experience one something terrible will happen such as a heart attack, or a brain haemorrhage, or that you'll go hysterical and make a total fool of yourself. If your problem is obsessive-compulsive disorder, then your fear may be that if you don't do things in the right order, or clean or check sufficiently, then something dreadful will befall you or those close to you. A key feature of post-traumatic stress disorder is that the person tries to avoid reminders of the trauma. They frequently think that if they're reminded too sharply of what happened, they'll start re-experiencing it, and that the feelings might be more than they can bear. In this chapter we will look at different anxiety disorders in turn. However, the techniques we discuss to manage anxiety are general ones. If your anxiety problem is more severe or specific then the further resources in Chapter 10 will help you discover where else you might get help.

If you are someone who feels anxious a lot of the time, or your anxiety is so intense it's starting to affect your everyday life, you may be suffering from one of the anxiety disorders. While we mentioned that anxiety is normal in certain situations, it becomes a problem when:

- It is out of proportion to the stressful situation

- It persists when a stressful situation has gone

- It appears for no apparent reason when there is no stressful situation.

USEFUL TIP: WHERE TO START WITH ANXIETY

1. Try to understand your symptoms

2. Talk things over with a friend, family member or health professional

3. Look at your lifestyle – consider cutting down or steering clear of alcohol, illicit drugs and even stimulants like caffeine

4. Apply some of the CBT techniques in this chapter.

It's quite common for people who are suffering from anxiety to also have symptoms of depression. If this is true for you then Chapter 6 on managing depression may be helpful for you.

CBT looks at how our thoughts, emotions, physical sensations and behaviours all interact to maintain our anxiety. When we perceive a 'threat' of any kind – whether that is a fear of something that is happening right now or a worry about something that might happen in the future – our bodies and minds react in the ways we look at in the diagram opposite. When we notice the physical sensations of anxiety we assume that this means there really is a threat (even if in reality there is none) and so we get more anxious thoughts. This in turn leads to

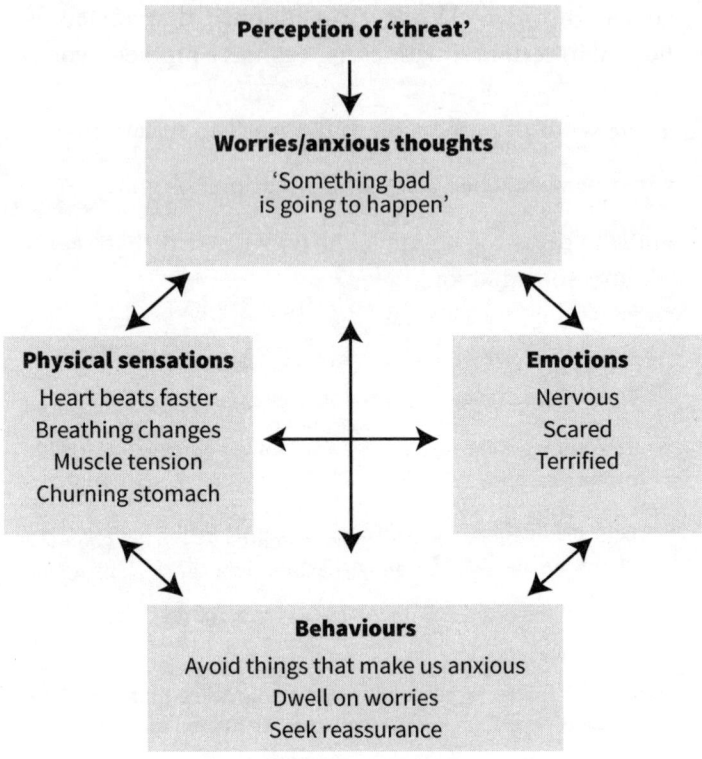

Perception of 'threat'

Worries/anxious thoughts
'Something bad
is going to happen'

Physical sensations
Heart beats faster
Breathing changes
Muscle tension
Churning stomach

Emotions
Nervous
Scared
Terrified

Behaviours
Avoid things that make us anxious
Dwell on worries
Seek reassurance

How CBT understands anxiety

enhanced physical sensations as our bodies respond to our perceptions. When we are scared of something we naturally avoid it. However this in turn can lead us to believe more strongly that there really is something to be scared of – and by avoiding it, we never get the chance to test out our fears. Our anxiety about that situation therefore increases. Often we dwell on our fears and worries in order to try to make sense of them, keep ourselves safe or stop bad things happening. However, this habit is most often unproductive and simply serves to increase our anxiety without actually improving or changing our situation. Seeking out reassurance from people close to us, searching the internet, or consulting professionals might make good sense if we do it once and it serves to calm our fear in a lasting way. However, what tends to happen when people suffer from anxiety is that they will seek reassurance, feel better for a short time, but then keep needing more reassurance. This means that nothing changes and they never develop more effective, lasting ways of managing their anxiety.

REMEMBER
If you are suffering from problems with anxiety, you are certainly not alone. Difficulties with anxiety are common within the general population. One in eight adults will suffer from an anxiety disorder at some point in their life.

There are several types of anxiety disorder – generalized anxiety disorder (GAD), panic disorder, agoraphobia, obsessive-compulsive disorder (OCD), phobias, post-traumatic stress disorder (PTSD), social anxiety disorder, health anxiety and stress reaction disorder. They all have some symptoms in common.

Listed below are the key areas which point to problems with anxiety. Do any of these describe you?

- Difficulty relaxing?
- Nervous, anxious or edgy?
- Easily annoyed or irritable?
- Restless and unable to settle?
- Unable to stop or control worrying?

- Worrying about practically everything?
- Fearing something awful might happen?

If any frequently apply to you, it may be useful to see your GP and talk through what's going on and what help is available, including of course self-help books like this one.

WHAT IS AN ANXIETY DISORDER?

Let's look in more detail at the different types of anxiety disorders. They all share many common elements. We will then explore the techniques CBT employs to help people deal with them.

GENERALIZED ANXIETY DISORDER (GAD)

Suffering from **GAD** means you'll be feeling anxious, tense and will worry most days, often about things other people consider quite minor. If you don't tackle it, the problem can last years, severely interfering with quality of life. Generalized anxiety can frequently be something that people feel they have always experienced to an extent – 'I've always been a bit of a worrier' – but which becomes more disabling during or following periods of increased or intense stress. Sometimes it can become more of a problem following distressing events such as bereavement, redundancy or a relationship break-up, and can start some considerable time after these events.

Women are more likely than men to be diagnosed with GAD, perhaps partly because women are more willing to see their doctor and admit to such feelings. You are more likely to experience GAD if you are aged

35–54, if you are divorced or separated, or if you are a single parent – but anyone can develop this problem.

Usually someone with GAD recognizes their worries are excessive and inappropriate. Sometimes, though, they aren't even aware of what it is they're worried about – they just feel uncomfortable and can't settle or relax. For a diagnosis of GAD you'll usually also have three or more of the following symptoms:

- Restlessness
- Irritability
- Tiredness
- Physical tension
- Disturbed sleep
- Problems concentrating or feeling as if your mind just goes blank.

case study   JANE (GAD)

Jane is in her early thirties. Her young child has just started school. Jane's back at work and wants to make a good impression on her new boss who was appointed during her maternity leave. She's always been a bit of a perfectionist, but previously had time available to devote extra hours to meeting her excessively high standards. Now with the additional demands of motherhood and work, she feels it is all too much. At work, she worries about not being as quick and efficient as colleagues who haven't had a maternity break. She also experiences anxiety about how her child is coping, feeling she should be a full-time mum, but knowing her income is required to make ends meet. There's no peace at home – work-related thoughts intrude constantly, as do self-critical thoughts about her ability both as a mother and a wife. As for GAD symptoms, she has a full house! Constant worry and restlessness, sleep problems, physical feelings of tension and various aches and pains.

POST-TRAUMATIC STRESS DISORDER

When people experience a trauma such as being involved in a car accident or being attacked or mugged it is very common for them to experience fear, recurrent and distressing thoughts and memories of the event, a sense of emotional numbness, a distance from those around them and intense anxiety. They may also try to avoid any reminders of the event or its consequences. These symptoms are all very normal and are part of a process of adjusting to and making sense of what has happened. Generally, these symptoms diminish in the few weeks following a trauma and most people recover well with time and support. For some people, however, these symptoms persist or even worsen over time and it feels impossible to move on from what has happened. In some cases symptoms can continue or even suddenly begin months or even years after the trauma. This is **post-traumatic stress disorder**. We discuss it and suggest ways of coping with it in Chapter 7, which covers using CBT to cope with difficult life events.

PHOBIAS

A **phobia** is a strong fear or dread which is out of proportion to the reality of the situation causing it. Coming near or actually in contact with the feared thing or situation causes anxiety, and just thinking of what you are phobic about is frightening and upsetting. You may sometimes be able to avoid the feared situation, but in many cases this can mean restricting your life. Also, the more you avoid, the more you may want to avoid and this can become more and more limiting over time.

There are many phobias of specific things or situations. Common examples include claustrophobia (fear of confined spaces or of being trapped), fears of specific animals and fears of injections, vomiting or choking. There are dozens of phobias, but the treatment for them all follows the same principles of graded exposure which we discuss later in this chapter.

SOCIAL PHOBIA

Social phobia or social anxiety disorder is possibly the most common phobia. You become very anxious about what other people may think of you, or how they may judge you. You fear meeting people, or 'performing' in front of others, especially strangers. You fear that you will act in an embarrassing way and that other people will think that you are stupid, inadequate, weak, foolish or even crazy. You avoid such situations as much as possible. Psyching yourself up to go somewhere is really hard – you often leave invitations open to the last minute, so as not to have to commit yourself. If you do go to the feared situation, you are often very anxious and distressed, and may well leave early. As with all anxiety disorders, the key to overcoming social phobia is to use a combination of thought challenging and behavioural experiments (see later in this chapter).

PANIC DISORDER

People with **panic disorder** experience recurring panic attacks. A panic attack is a severe attack of anxiety and fear which occurs suddenly, often without warning, and for no apparent reason. The physical symptoms of anxiety during a panic attack can be severe and may include: a thumping heart, trembling, feeling short of breath, chest pains, feeling faint, numbness or pins and needles. Each panic attack usually lasts 5–10 minutes but sometimes they come in waves for up to 2 hours. Panic attacks are incredibly frightening experiences and whilst they are happening people can really feel as if they are dying. This naturally leads to fear of fear – feeling scared that an attack will occur and that *this* time it will finally be the one where something terrible *really* does happen.

People often try to cope by avoidance, shying away from any situation in which they think an attack might happen or where they might not be able to escape from the panic. This can severely limit some-one's life and for some people is also associated with agoraphobia.

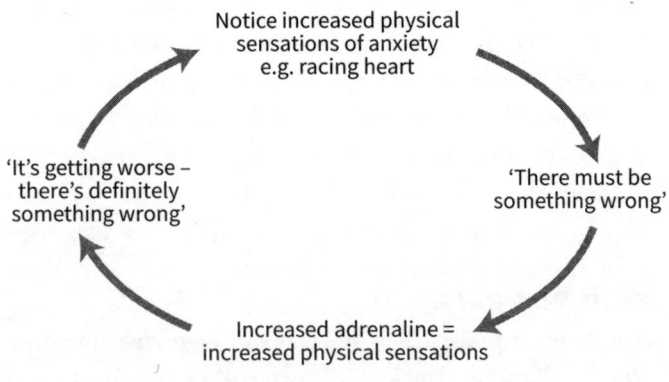

The panic cycle

AGORAPHOBIA

Agoraphobia in ancient Greek literally means 'fear of the market place'. The term describes a fear of open spaces and frequently includes difficulties being in public places – shops, crowds, on public transport, crossing bridges or even simply being away from home. It is usually difficult, if not impossible, to do these things alone, though some sufferers of agoraphobia may manage to go out and about if accompanied by someone they trust.

All the different situations which cause difficulties for people with agoraphobia are united by one underlying fear – that of being in a place where you are overwhelmed by panic, no help is available, and you'll find it difficult if not impossible to escape to a safe place (usually to your home). When you are in a feared place you become very anxious and distressed and have an intense desire to escape. To avoid this anxiety and panic many people with agoraphobia stay inside their home for most or all of their time. Sadly, however, they can then experience panic attacks even in their home and so feel they have to have someone with them at all times.

Agoraphobia and panic disorder affect around 5 per cent of the population, affect women more than men, and most commonly occur between the ages of 25 and 35. Agoraphobia affects up to one third of people with panic disorder and occurs before the onset of an attack. The fear means that the person tries to avoid places where they are likely to have panic attacks and, while avoidance can be successful to some degree in keeping panic attacks at bay, the restrictions on a person's life usually just keep increasing, affecting both the person and those close to them.

case study
BILLY (PANIC DISORDER WITH AGORAPHOBIA)

Billy is a 25-year-old office-worker. He commutes daily using public transport. He used to like the train journey. It gave him time to read the paper and relax a little before the start of a stressful day.

One day, however, the train was particularly crowded. It was very hot and the train's air conditioning had failed. Billy started to feel very warm. He noticed that he was sweating and that his heart had started to race. His chest hurt and he felt shaky. He thought something must be very wrong. He was convinced he was having a heart attack. He got off the train at the next stop and called an ambulance. In A&E he was examined and told that there was nothing wrong with his heart. He had had a panic attack. He felt very relieved but shaken and frightened by what had happened. He had really felt like he was dying. He never wanted to feel like that again.

The next time he went on the train he was very anxious and again started to notice symptoms. Again he focused on these and experienced the frightening feeling that he was dying. The sensations were very difficult to cope with and he was forced to get off the train and go home. Gradually, Billy found that his fear of the panic symptoms led to him avoiding more and more situations where he thought they might occur and where he was afraid he would not be able to escape.

Over time, Billy's avoidance became more and more entrenched. He did not believe he could withstand or manage the panic symptoms and so he would simply not do anything that he felt might trigger them. He gave up his job and started to work from home. Slowly he went out less and less and his social life dwindled. As he avoided more situations, so his fear that a catastrophic attack would happen increased and he felt increasingly unable to go out at all.

OBSESSIVE-COMPULSIVE DISORDER (OCD)

Obsessive-compulsive disorder (OCD) consists of recurring obsessions, compulsions or both.

Obsessions are recurring intrusive, uninvited and unwanted thoughts, images or urges that cause you anxiety or disgust. Common obsessions are fears of being contaminated by dirt, germs, disease or body fluids, and also fears of disasters. They can encompass worries about violence that will happen to you, or harm you might do to others despite it being against your will, including paedophilia and bestiality. Fears related to religious beliefs are also common.

Compulsions are thoughts or actions that you feel you *must* force yourself to have or do, and often that you feel you have to keep repeating, until you have got it 'right'. Usually a compulsion is a response the person makes to ease the anxiety caused by an obsession. A common example is repeated hand washing in response to obsessive fear of dirt or germs. An individual may disproportionately fear that they have dangerous germs on their hands from touching things and that these could be harmful to themselves or people around them. They may therefore feel a compulsion to very frequently wash or disinfect their hands in order to reduce this fear. Other examples of compulsions include: repeated cleaning, checking, counting, touching, placing objects in particular positions and also hoarding objects.

Often professional help is required, as it can initially be quite difficult for a person to discriminate between obsessional thoughts and actual danger. Likewise, if you fear you are going to harm someone against your will, understandably you won't be keen to put this to the test, just in case you were to find out it was indeed so.

People suffering from OCD often have an exaggerated sense of responsibility. They may feel it is their role to protect themselves from the dangers of the world, the threats of which they usually considerably overestimate. They may also feel they must ensure that harm does not come to others. Very often carrying out the ritual or compulsion still doesn't solve the problem. They might experience intrusive thoughts about harm coming to others. So, for example, they may feel they have to move stones off pavements to avoid someone tripping but then go on to worry that the new place they have moved them to could cause harm to someone instead.

HEALTH ANXIETY

Some concern about your health can be useful, as it means you may try to lead a healthier lifestyle. People who have had health problems, in particular something like a heart attack or cancer, often decide to take them as a warning that unless they make certain changes, then something worse could happen next time. While that attitude can be very productive for some, others find they become increasingly obsessed with their health. Some people find this increased anxiety happens to them following the illness of someone they know, after an important life change or just out of the blue for no obvious reason at all. Any minor symptom is blown out of all proportion.

A minor sniffle equates to imminent death from swine-flu, a mark on the skin means malignant cancer, tiredness is multiple sclerosis, while a headache equates to a brain tumour – which will, of course, be inoperable. People with health anxiety visit their doctor frequently and can end up having many investigations, tests and visits to specialists which often come to nothing. They may also spend a large amount of time researching illness on the internet and in books. The worry and fear of illness can take over people's lives and cause considerable misery.

case study MAMTA (HEALTH ANXIETY)

Mamta is a fit, healthy woman in her 50s. Tragically, one of her closest friends recently died of breast cancer. She had been a very healthy woman who took good care of herself, ate well and exercised regularly. Her cancer came out of the blue. She went through many months of distressing treatment before dying at the age of 53. Mamta is naturally very upset by this loss and also frightened by the way in which her friend suddenly became ill.

She begins focusing on her own body and on anything she experiences which could be interpreted as a symptom of ill health. If she has a headache or a muscle twitch, or notices an ache anywhere in her body she becomes totally preoccupied with this, worrying endlessly about what it might mean. Mamta looks symptoms up on the internet, or asks others what they think is wrong with her. She visits her GP more frequently, asking for reassurance that what she sees as being symptoms aren't signs of something serious. Having seen the doctor, Mamta feels better for a short while but then starts worrying again. Her GP sends her for several tests because Mamta is worrying about symptoms which previously she'd have ignored.

As Mamta's anxiety and stress increase, she experiences more physical symptoms related to anxiety – more frequent head-aches, fatigue, palpitations and general aches and pains. These of course add to her worries and send her back to the GP and to other sources of reassurance. When assured that there's nothing wrong, Mamta begins to doubt her doctor and starts exploring an ever-increasing range of different types of therapies.

HOW CBT CAN HELP WITH ANXIETY

The good news is that there are lots of tried and tested techniques which have been developed by CBT therapists to help people overcome anxiety disorders. The following methods can all be applied to a variety of the disorders described above. Some will be more helpful in certain situations than others. Try them out and see which work for you. We will first look at how you might deal with your thoughts and will then move on to examine some behavioural strategies which may be helpful to manage your anxiety.

THOUGHT BALANCING

This technique is key to cognitive therapy and involves looking at your anxious thoughts in a different way. You really can start to see that thoughts are just that, and do not necessarily represent facts. We can see thoughts as simply **mental events**. Yet frequently we respond to them as if they were concrete facts rather than possibilities or ideas. Just because you think something bad will happen is that guaranteed? Some thoughts will be true, some won't and there will be a large grey area in between these two extremes. Remembering this is a good place to start.

Chapter 6 looks at depression and shows you how to begin to manage your negative thoughts by examining the objective evidence for them and working on developing alternative balanced (and ultimately more realistic and helpful) ways of viewing a situation. Check out some of the exercises there. Consider whether you are being affected by any of the distorted ways of thinking that we describe.

try it now
CHALLENGE YOUR WORRIES OR ANXIOUS THOUGHTS

When dealing specifically with anxious thoughts, worries or predictions it can be helpful to ask yourself the following questions to help you gain a different, and potentially more helpful, perspective. Write down the answers for each of your separate worries or thoughts.

- How important will this be in my life 5 years from now?

- What would my best friend say I should do about it?

- What would I advise my best friend to do if this was their problem?

- Am I assuming my way of seeing things is the only one possible?

- Am I jumping three events ahead when the first step hasn't even happened yet?

- Am I overestimating the chances of disaster?

Now think of some of your own questions to challenge your worried, anxious thoughts, and write them down. Feel free to write as many as you like!

CHALLENGE PERFECTIONISM

Do you always expect more of yourself than it is possible to achieve? Do you have much higher standards for judging yourself than for other people? If this is true then you may be falling into self-defeating patterns that are maintaining your anxiety and making you very unhappy. Look at the section in Chapter 6 which discusses how to reduce self-criticism.

Watch out for 'should', 'always' and 'must' in the way you talk to yourself – these words are rarely helpful. Jane from our case study frequently says to herself 'I *should* be doing this perfectly', and 'I *must* get all this right first time all by myself or I'll look a right idiot!' These statements just make her feel miserable.

Try to stick to the rule 'good enough is good enough'. You can still be good at things, but aiming for excellence rather than perfection is much more likely to get you results. Do you know anyone who is actually perfect in every way? No? So how come you expect yourself to be able to achieve this?

EXAMINE YOUR BELIEFS ABOUT WORRY

Lots of people find worry a problem because of the beliefs they hold about the process of worry. Some people have positive beliefs about worry such as:

- Worrying helps stops bad things happening and helps me stay safe
- Worrying helps me be more organized
- If I didn't worry constantly, I'd always get things wrong.

At the same time they may also have negative beliefs about worry such as:

- Worrying could drive me crazy
- Worrying could make me ill
- Worrying puts a strain on my heart.

With beliefs like these no wonder some people find it hard not to worry – and then feel very afraid when they can't stop. Think about your own worry beliefs. Write them down in the table below.

Why worrying is useful to me	Why worrying is bad for me

Now try to think them through logically. When we worry, we are actively trying to anticipate every bad thing that could possibly happen in order to (somehow) prevent it. But is this actually possible? Isn't it true that sometimes unfortunate things just happen no matter how much we have thought about things beforehand? Can we really cause things to happen or not happen with our thoughts?

Most worry is unproductive. Time we spend on it doesn't help us to be more organized or effective. The outcome of worry is usually just more stress, tension and anxiety. Research shows that stress can affect us physically. However there is very little evidence that stress and worry alone (in the absence of pre-existing medical problems) can actually cause lasting or catastrophic damage to us physically. Worry is unlikely to harm either your physical or mental health. However, what we can guarantee is that worry will cause you to be more miserable and enjoy life less – all the more reason to work on dropping it.

REMEMBER
The beliefs we have about worry can cause us more distress. Examine and question these beliefs – are they really true?

USE THE WORRY DECISION TREE
In the table you've probably written down quite a few worries about your worrying. Are you now worrying what to do about it? Fear not, an exercise is at hand. Let's deal with our worries by having a worry plan. When you notice yourself worrying, go through the following exercise.

try it now

Ask yourself: What am I worrying about? Write down each worry separately. Keep going until you have listed them all. Then for each separate worry go through the following diagram.

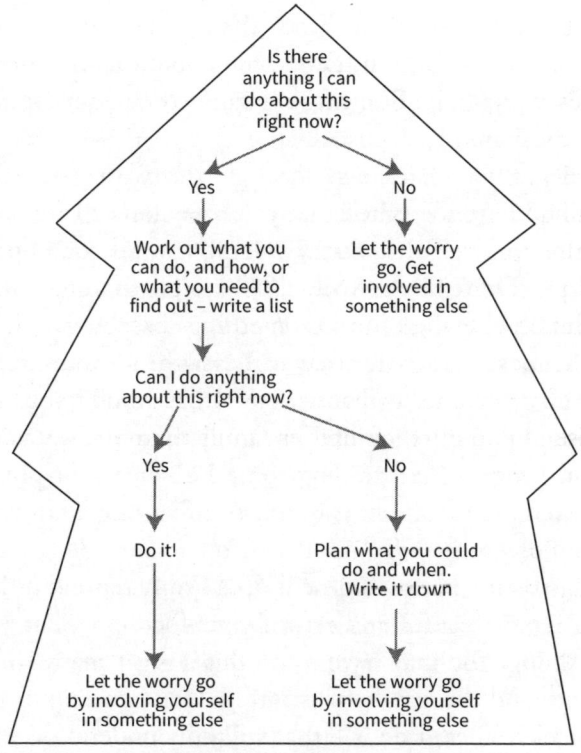

Is there anything I can do about this right now?

Yes → Work out what you can do, and how, or what you need to find out – write a list

No → Let the worry go. Get involved in something else

Can I do anything about this right now?

Yes → Do it!

No → Plan what you could do and when. Write it down

Let the worry go by involving yourself in something else

Let the worry go by involving yourself in something else

The worry decision tree

ABSORBING YOURSELF IN SOMETHING ELSE

It's very easy for psychologists to tell you to 'distract yourself' when you are worrying or in the middle of a panic attack – but very difficult to do this in practice!

Frustration with this is very natural. Our minds are very busy places – they are designed that way. We experience countless thoughts in a day and emotionally charged ones like our worries can be very hard to ignore. However, we know that with practice and patience you can learn to move on rather than remain stuck with them. Telling yourself 'Don't think about it!' most certainly won't help. Be firm (but kind) with yourself and your busy mind.

Once you have run through the worry decision tree and have identified that you have done all you can, remind yourself that further worry will just be unproductive. Don't berate yourself – just gently direct your attention elsewhere, on to something absorbing.

Choose an activity that will easily hold your attention and focus all of your senses on it. Maybe you will choose talking to a friend or family member, watching a TV programme or doing some housework or physical exercise. Whatever it is, practice focusing all of your attention on it. Your mind will try to intrude on this with worries, but each time it does firmly remind it that this is not helpful and return your focus to what you are doing. You may have to do this a great many times at first and this can be frustrating. Don't give up or tell yourself you can't do it – that will only undermine your good work. Nobody gets this straight away. It will take a lot of practice but over time it will become easier to do.

REMEMBER
Getting involved in alternative activity isn't easy. Patience and practice is the key.

AND NOW ... RELAX ...

Believe it or not, being able to physically relax is quite a skill – one which unfortunately many of us have never learnt adequately. When we are busy rushing about from one task to another, day after day, we can very often find that we carry a lot of tension in our muscles. Do you ever find your shoulders, neck or back aching by the end of a long or stressful day? Much of this may be due to muscle tension. When we are stressed, worried or anxious, this tension increases and can result in aches and pains, headaches and fatigue. Spending some time learning to relax physically can be a worthwhile investment to help you to cope better with anxiety or stress. Like learning any skill it takes practice. The following exercises should be repeated daily when you are learning. Setting some time aside each day to carry out an exercise like this will also help you develop the good habit of prioritizing a short time for daily relaxation. Once you know how to relax, keep up this habit – finding a time each day to help yourself relax physically can make a big difference.

try it now EXERCISE 1: DEEP MUSCLE RELAXATION

- Select a place that is warm and comfortable, where you won't be disturbed. Initially, choose a time of day when you are likely to be feeling most relaxed.

- Lie down, get comfortable and close your eyes.

- Concentrate on your breathing for a few minutes. Breathing slowly and calmly, count 'in-two-three, out-two-three'.

- You will now work through different muscle groups, teaching yourself first to tense, then to relax. You should breathe in whilst tensing and breathe out when you relax.

- Start with your hands. First clench one tightly. Think about the tension this produces in the muscles of your hand and forearm.

- Study the tension for a few seconds and then relax your hand. Notice the difference between the tension and the relaxation. You might feel a slight tingling. This is the relaxation beginning to develop.

- Now do the same with the other hand.

- Each time you relax a group of muscles think how they feel when they are relaxed. Don't try to relax, just let go of the tension. Allow your muscles to relax as much as they can.

- Focus on the difference in the way they feel when they are relaxed and when they are tense.

- Now do the same for the other muscle groups in your body. Each time, tense them for a few seconds and then relax. Study the way they feel and then let go of the tension in them. It is useful to keep the same order as you work through the muscle groups ...

- **Hands** – clench first, then relax.

- **Arms** – bend your elbows and tense your arms. Feel the tension especially in your upper arms. Remember, do this for a few seconds then relax.

- **Neck** – press your head back and roll it from side to side slowly. Feel how the tension moves. Then bring your head forward into a comfortable position.

- **Face** – there are several muscles here, but just concentrate on your forehead and jaw. First lower your eyebrows into a frown. Relax your forehead. You can also raise your eyebrows, and then relax. Now clench your jaw, then relax – notice the difference.

- **Chest** – take a deep breath, hold it for a few seconds, notice the tension, then relax. Let your breathing return to normal.

- **Stomach** – tense your stomach muscles as tight as you can … and relax.

- **Buttocks** – squeeze your buttocks together, then relax.

- **Legs** – straighten your legs and bend your feet towards your face. Relax. Finish by wiggling your toes.

You may find it helpful to get a friend to read the instructions to you. As you go through the exercise, don't try too hard – just let it happen.

try it now
EXERCISE 2: A NEW TAKE ON 'BIBLIOTHERAPY'

- This is not reading a book – it's using one! You'll need just 15 minutes, though if you want to do it for longer, enjoy!

- Find a reasonably large book, take it to a quiet place and then set an alarm for 15 minutes, so you needn't worry about the time.

- Lie down, open the book and put it face down on your tummy.

- Concentrate on breathing in slowly through your nose for a count of 4 seconds, hold your breath for 2 slow seconds, then exhale slowly through your mouth for 4 more seconds. Keep repeating this step.

- Focus your attention on the book. Watch it rise and fall. Study it as closely as you can.

- Your busy mind is bound to try to intrude with lots of distracting thoughts. Don't follow through with them, but rather tell yourself you'll deal with them later, as you are currently doing your bibliotherapy. Then back you go to concentrating on counting, breathing, and watching the book slowly rise and fall, rise and fall … is that the alarm already?

try it now EXERCISE 3: A SAFE PLACE

This is a visualization exercise which can take some practice to get the hang of. It's also called **self-hypnosis**. People often think they are not good at visualization, but with practice and patience most of us can actually conjure up pictures in our minds. What's great news is that research shows you don't need vivid pictures for this to work – faint, fuzzy ones are equally good.

- Find a quiet space and sit or lie comfortably. Again, it might help to set an alarm for 15 minutes so you aren't worrying about the time.

- Relax and concentrate on your breathing. Breathe in and out slowly and deeply from your stomach. Aim to slow your breathing down to 10–12 breaths per minute, but then forget about it. Just breathe naturally.

- Close your eyes and start to imagine yourself in a place that's safe and warm and peaceful. This could be anywhere – a tropical beach, a park in summer, your bed or the moon. It can be real or imaginary.

- Focus on your senses. What can you see, hear, smell, taste and touch in this safe place? How do you feel while you are in this place? What's around you?

- Spend a few minutes exploring your safe place. Relax your muscles and let all the tension disappear while you are in your safe place.

- Again, your busy mind may try to distract you with other thoughts, worries or images. Just gently let them go. Remind yourself that right now you are in your safe place – the other thoughts can be dealt with later. Practice **turning down your thoughts** – just like you turn down the volume on a radio.

With practice, you'll find that you can very usefully call up your safe place whenever you are stressed or anxious. Going to this place briefly in your mind can help you refocus, calm down and then be able to move forward in a less stressed state of mind.

BREAK THE PANIC CYCLE

Earlier we looked at how thoughts and physical sensations interact to create a panic attack. The first step in dealing with panic attacks is to educate yourself about what causes them. Panic feels incredibly awful – terrifying. It's very hard to believe that something catastrophic isn't happening to you. However, we know that panic is a self-limiting system. It cannot harm you. There is no evidence that anyone has ever died from having a panic attack without having an underlying health condition. Neither is there evidence that anyone has ever 'gone insane' from having a panic attack. Just because you are feeling very strong physical sensations does not mean that the catastrophe you believe will happen is inevitable. The chances are it won't happen at all.

For example, many people feel that when they have a panic attack they'll pass out. But guess what needs to happen to your blood pressure for you to pass out? It needs to drop suddenly. What do you think generally happens to your blood pressure in a panic attack? It increases (though generally not dangerously). The only exception to this is if you have a phobia of blood or injury, in which case seeing these things could make your blood pressure fall. So, it's virtually impossible for most people to pass out during a panic attack.

Consider what your fears are when you panic:

- What is the worst thing that could happen?

- How likely is it that this will actually happen (rather than how much it feels like it will)?

- How likely would someone else consider it to be?

- If the worst did happen, how likely is it that you would not be able to cope with it (no matter how awful it was)?

Thinking in this way can be very helpful when trying to break a panic cycle. However, ultimately the only way you will *prove* to yourself that all this is true is by facing your fears and testing out this new way of looking at things. The sections below on **graded exposure** and **behavioural experiments** will help you do this.

GRADED EXPOSURE

Exposure therapy is the way that CBT helps people to face up to and overcome their fears. It is used in various ways in treating all the anxiety disorders. Here we describe the fundamental principles that apply to them all.

1. **Develop a graded hierarchy.** Write the numbers 1 to 10 along the side of a piece of paper. 1 represents activities that aren't particularly scary – those you would be a little anxious about but could do by pushing yourself. 10 represents activities that hold the most fear for you – the ones that make you say 'No way! I could never do that!' Start by writing

something down for the top and bottom of this scale. Then think of what might go in the middle – what activity would rate about a 5? Continue until you've got 10 activities, and have filled out the scale. Getting help from someone who knows you well can be invaluable.

2. **Start working through your hierarchy from the bottom up.** You may ask someone close to you to accompany you if that helps to get you started, but then it is very important that you continue to practice on your own. Keep repeating each item until you begin to feel confident about it, and until you find that the scary outcome you feared, whatever it might be (spiders crawling all over you, bats getting tangled in your hair, falling from a height or even drowning), doesn't happen. At each stage make a note of what you learned and use that to help you progress to the next step on the hierarchy. Do remember to congratulate yourself each step of the way, rather than saying 'it's easy for others so my achievements are no big deal'. These are big achievements *for you*.

3. **Stay in the situation until your anxiety falls.** At the beginning of each exposure task, rate your anxiety out of 10. It is likely to be very high at first. It is very important that you stay in the situation until your anxiety falls to at least half of what it was originally. This can be tough. However, if you 'escape' too early you won't learn how you *can* cope with your fear. Stay put – your anxiety will fall and you will learn what you need to in order to progress to the next step.

A WORD ABOUT SAFETY BEHAVIOURS

Think about the following story:

A man was sitting outside, rhythmically clapping his hands. A boy approached him and asked 'Why are you clapping like that?' The man replied, 'It keeps the tigers away'. The boy smirked and looked around, 'But there aren't any tigers!' And the man replied, 'See? It works!'

Why is this story amusing? The man really believes that his clapping is keeping tigers away. He believes this because while he does it there are no tigers. However, what he doesn't know, but the boy does, is that there are no tigers in the first place. The only way for the man to discover this would be to stop clapping and test out his belief – but this is very scary to do if he really believes the clapping is keeping tigers away. The clapping is what we call in CBT a **safety behaviour**. People with anxiety frequently use safety behaviours to try to help them cope better. For example, they may:

- Carry a bottle of water when they are out in case they get hot and start to panic

- Sit close to the door on a train so that they can escape if something bad happens

- Try to control their breathing

- Use headphones to block out the sounds of other people talking (if those sounds increase their anxiety)

- Bury their head in a newspaper if they imagine people are looking at them.

All of these things may help them to tackle things they wouldn't otherwise do. The problem is that, like the man clapping they continue to believe that the feared outcome – the 'something bad' – would definitely have happened if they had not carried the water, used the headphones, sat by the door, controlled their breathing and all the rest. In this way these behaviours prevent the person from really testing out if they can cope with their fear, and finding out if the thing they are afraid of really happens. So, the anxiety never really goes away.

While you are doing your exposure work, watch out for safety behaviours. If you need to use one to do the exposure at first, that's fine. But remember it's a safety behaviour. You need to drop it as soon as possible and carry out the exposure without using this crutch – only then will you really conquer your fear.

BEHAVIOURAL EXPERIMENTS – STEPPING UP THE PACE

Taking things a step beyond graded exposure, you can become a scientist, testing out your anxious thoughts and fears. A scientist designs experiments to test out theories and hypotheses about the way the world works. In CBT we do the same thing.

STEP 1

Think of a situation which you avoid because you are afraid something bad will happen. What is it you are afraid of? What is the worst thing that could happen? What is your most anxious prediction about what might happen if you were to put yourself in this situation? Write down this fear or prediction. Remember that your prediction should not simply be that you will get anxious in the situation – we already know that to be true. There would be no point in doing this experiment if the situation didn't make you anxious. Your fear is likely to be more than that – what is it that you think will be the consequences of getting anxious? Losing control? Not coping? Falling apart? Becoming dangerously ill? Making a fool of yourself? Use your imagination – what are you really terrified will happen?

STEP 2

Design an experiment to test out that prediction. What do you need to do? How would you measure whether the prediction was true or not? Write all this down. Rate out of 10 how much you believe your prediction will come true. Also think about what might stop you carrying out your experiment – how could you overcome such obstacles to ensure you do complete it?

STEP 3

Carry out the experiment. Remember to use whatever way you have decided on to measure what happens. Write down what happens.

STEP 4

Okay. What did happen? Did your most feared prediction come true? What did you learn? Write all this down – this can then help you to design your next experiment.

case study
BILLY (PANIC DISORDER WITH AGORAPHOBIA)

Remember Billy, our panic and agoraphobia case study? Here is an experiment that Billy did to help test out his fears, and what he learned from it.

Step 1
Anxious Prediction: If I go to the local shop I'll have a panic attack and I won't cope with it. I'll faint or lose control in some way and make a total fool of myself. I believe 80% that I'll lose control if I get anxious.

Step 2
Experiment: To walk to the local shop and go inside. To spend a few minutes looking at the magazines. To stay for at least 5 minutes and then to come home.

What might stop me? I might get overwhelmed with fear and not be able to go through with it.

How can I overcome this? I'll write down the rationale for this experiment and use it to remind me why I am doing this. Remembering this will help. I'll get a friend to encourage me to leave the house. I'll arrange to do something nice afterwards as a reward.

Step 3

What happened? I did it! It was really tough and I did feel pretty bad. My heart raced and I felt very wobbly. I was exhausted afterwards. But I didn't freak out or pass out and I don't think anyone really noticed how awful I was feeling.

Step 4

What did I learn? That although I feel awful it is not as terrible as I thought. People don't seem to notice my anxiety as much as I think they will – perhaps it is not as obvious as I assumed. I do panic a bit, but I don't lose control. I now believe only 40% that I will lose control if I get anxious.

What next? I'm going to try this in different situations – test out what I fear.

ACTIVITY SCHEDULING AND PLANNING

Very often we feel anxious and panicky because we have taken on too much or not planned our time effectively. Effective planning is a very important life skill which many of us (including Elaine and Clair, your authors) need to work on.

When we have a lot on, we often get anxious and the anxiety can paralyse us – stopping us being able to tackle the many things we have to do. There are a few rules which we can follow to prevent us from getting overwhelmed and stuck in this way.

1. BE MORE LIKE A HUMMINGBIRD THAN A BUTTERFLY

Watch a butterfly. It seems to flit from one place to the next, and when it stops you can't really see it doing anything before it flits off again. When we are anxious we tend to behave like that butterfly – dashing from one task to the next, trying to do too much at once and ultimately not doing anything properly or completely. A hummingbird, on the other hand stays in one spot, hovering despite the pull of gravity, drinking the nectar out of one flower before proceeding to the next. The rule is, however much you have to do, **do one thing at a time** and focus on just that one thing until it is finished and you can move on.

2. BREAK THINGS DOWN INTO MANAGEABLE STEPS

Have you ever looked at all you have to do and felt overwhelmed, not knowing where to start? It's so tempting to abandon any attempt to start your task and to just bury your head in the sand. Instead, **break tasks down into small steps**. What's the first thing you need to do? Then just do that first step without worrying about the next. Now move on to the next small step and before you know it you will have completed what felt like a mammoth task. Use the 5-minute rule we discuss in Chapter 6. If something feels too overwhelming, just do it for five minutes. Don't think any further ahead than that.

3. WRITE OUT AN ACTIVITY PLAN

Each day, list out the tasks you intend to do. **Ensure your choices are realistic** (this probably means crossing out a few), then prioritize them. Which absolutely have to be done today? Which could wait a little? Decide what you want to do, when, and how much time you need for each task. Then add a bit of extra time for good measure. Draw up a timetable for the day. Build in 'bio breaks' (coffee and tea, meals and toilet stops), and even allow for brief day-dreaming periods! Then follow your timetable. As you work your way through, visualize yourself as that hummingbird, hovering away steadily until the task is complete, then heading for the next one.

4. PROBLEM SOLVE

If you are not sure how to deal with a certain task, **take some time to work it out instead of panicking**. Are there any sources of support you could use? Who could help? There's no shame in asking for help if you are stuck – how else would anyone learn anything? What would someone else say about this?

Write down the problem clearly. Now spend some time brainstorming all the possible solutions there might be. Really go for it – imagine as many as possible. Write them all down. Now go through each solution. Identify the pros and cons. Write them down. Give each solution a mark out of 10 after you have balanced the pros and cons. Then select the solution with the best scores. You could even get a friend to help you with this task. Not every problem will have an instant solution but breaking things down into small steps can often help us to see what we can do first or identify what information we need to gather to be able to come up with a solution.

Finally, test out the solution. Did it work? If not, why not? Go back to your problem solving with the new information and try something else.

REMEMBER
Learning to say 'no' can be very important. Do one thing at a time and try to avoid taking on too much in the first place.

MEDICATION

There are certain medications available, both on prescription and over the counter, which can help anxiety in the short-term. Beta-blocker medicine can ease anxiety and some physical symptoms such as trembling. It can be helpful for certain situational anxiety, like a performer wanting to reduce symptoms of shakiness before a concert. Beta-blockers are not addictive, are not tranquilizers, and do not cause drowsiness or affect performance, so you can take them as required. Sometimes your GP might prescribe diazepam (a benzodiazepine) as a short course for 2–4 weeks, if the cause of the stress is likely to last a short time, and if the symptoms are particularly acute and severe. You are unlikely to be prescribed diazepam for longer given the potential problem of addiction.

However, a note of caution: there is good research evidence suggesting that using medication alone to deal with anxiety doesn't prevent anxiety recurring in the future. Learning new ways to cope is usually very helpful, as the chances are the anxiety will return at times. There is also evidence that in certain cases using medication whilst undergoing CBT can in fact reduce the effectiveness of the therapy. Why should this be? The theory is that in order to learn to cope with anxiety and panic, you have to actually *experience* those feelings and develop ways to overcome them. Medication reduces the experience of anxiety in the short-term and so can prevent effective learning taking place. The only way to truly conquer anxiety is to learn strategies to manage it.

FINALLY ...

Let's see how putting into practice some of the things in this chapter helped our three case studies:

case study JANE (GAD)

Jane starts to put into practice the idea that 'good enough is good enough'. She keeps to her paid hours, and when she's unclear how to prioritize her work (too much to do in the available time) she asks her boss, and is told what she can drop. To her amazement, she becomes much faster at her work, and nothing's returned with errors identified. Her confidence increases, and within 6 months she's offered a promotion. Jane discusses it with family and friends, concluding that her present level of stress is about right. She tells her manager that she's keen to reassess the option in 6 months time, and this is agreed. At home, her sleep improves; she has more energy, and is less snappy with the family. When Jane has thoughts about not being good enough, she puts into practice thought balancing techniques, finding to her surprise that her self-esteem dramatically improves. The family start going on fortnightly outings, sometimes just to the local park, and all feel closer to each other. Jane and her husband also put aside time for themselves, as well as time together, and both are much happier with themselves and with their relationship.

case study
BILLY (PANIC DISORDER WITH AGORAPHOBIA)

As we have seen, Billy makes really good use of behavioural experiments to start testing out the fears he has about going out. He also reads about panic attacks and now understands more about the interaction between his thoughts, feelings, physical sensations and behaviours. He finds that once he recognizes what's happening, it's easier for him to control the panic. He still experiences it, but it gradually becomes less intense and he's less frightened that he'll completely lose control. He gradually starts to go out more and more. He makes arrangements with friends and asks them to support him to carry them through. Billy designs behavioural experiments to get back into using public transport and travelling on his own by train. He notices the safety behaviours he's using, such as sitting close to the door, and gradually reduces these so he really tests out whether the things he fears actually happen. He discovers that they don't. He is now seeking work outside of his home again and is enjoying his social life.

case study MAMTA (HEALTH ANXIETY)

Mamta understands the role that focusing on symptoms and seeking reassurance is having in maintaining her health anxiety. She enlists the help of her family and her GP to no longer provide her with reassurance but to encourage her to challenge her fears herself. She writes out a plan to cope when she notices what she regards as symptoms. Mamta now weighs up the evidence that the apparent symptoms are actually just passing, normal sensations. She postpones thinking about them and gets involved in alternative activities, as well as using relaxation to move her mind away from them. Mamta uses thought balancing to help reduce her fear. She promises herself to only go to the doctor if something she regards as a symptom persists for longer than a week. Gradually Mamta begins to feel better. As her anxiety and physical tension decrease, she notices that in fact she experiences far fewer 'symptoms'. She asks others about their physical sensations and is amazed to find that everyone occasionally experiences the things she was worrying about. The only difference is that they consider this to be normal. Mamta talks with her family about her grief over losing her friend. She still fears getting cancer, but slowly finds it no longer dominates her life and that she can get back to normal. However, Mamta regularly does a monthly breast check, and always goes for screening tests when invited by her GP.

Facing up to your fears can be one of the hardest things you'll ever do. Remember these words:

The greatest victory of all is victory over oneself.

CHAPTER 4

Kiss insomnia goodnight – and goodbye

*Nothing cures insomnia like the
realization it's time to get up!*

Kiss insomnia? You're kidding, aren't you? Surely this is asking too much! Bestow affection on insomnia, which may be the bane of your life?

What we will be showing you in this chapter is how to embrace the upsetting experiences that result from poor sleep. You'll then discover that when you learn to manage, tolerate and even embrace them, they usually happen with less frequency and severity. You may also discover that while disturbed, poor-quality sleep isn't much fun, it's all the other things that go with it that make it such an awful experience.

If you suffer from insomnia you aren't alone. It affects around one third of the UK and US populations, who have difficulty falling and staying asleep, wake up early in the morning or wake up feeling unrefreshed despite having slept. Women are nearly twice as likely to have it. It's more likely with age, often runs in families and is more common if you are a worrier or highly excitable.

The upsetting experiences that frequently accompany insomnia include daytime tiredness, problems concentrating, a greater chance of developing depression, anxiety and pain, a greater need for health-care services and a higher risk of car- and work-related accidents. Your intimate relationships can be affected – sharing a bed is less appetizing when you are desperate to sleep and fear another person disturbing what little sleep you may get. So, it's not just the lack of sleep but rather the resultant thoughts and physical sensations that seem to be at the very heart of the problem. Understandably, you want to be rid of them. But read on and you may find yourself surprised at what recent research has suggested; the new guidance could be exactly the opposite of what you expected and thought you needed.

However, before we look at changing anything, it's useful to understand sleep – why do we sleep, and how much sleep do we need? While most people will agree on wanting more sleep (just 5 minutes more!), how much sleep we need varies considerably. Some adults thrive on 5 hours a night, others say they need 8 hours' sleep. Our need for sleep changes with age, so babies can sleep up to 16 hours, while elderly people often need less sleep than in their youth.

We sleep in cycles of different depths and types of sleep, of around 90 minutes per cycle. A small awakening between each cycle means you can change position to avoid stiffness or go to the toilet. Normally we average 4 to 5 cycles a night. Frequent awakenings improved our chances of survival in years gone by, as we could check for dangers like wild animals or warring tribes. In today's world we need to train our brains that it's safe to go back to sleep after each cycle. For those with insomnia, normal waking between cycles can bring scary thoughts about the consequences of not sleeping, as well as anxiety and frustration, which can make you struggle to get back to sleep. The process of struggling feels like fighting a danger – which makes you wide awake!

About 50 per cent of our sleep is **light sleep** which feels like dozing; if woken during light sleep, people often believe they haven't been sleeping. **Deep sleep** happens usually in the first 2 cycles, and accounts for around 20 per cent of your sleep. This is when your body grows and repairs itself. Waking during this period, in the day or night, you'll probably feel confused and disorientated, as you have been in such a deep sleep. In the late night and very early morning you have the longest period of **REM (rapid eye movement)** sleep, which constitutes around 30 per cent of total sleep. Your brain is very active during this phase, laying down memory, going through emotions and dreaming. Waking from this stage can also mean feeling tired and unrefreshed.

Falling asleep depends partly on something called the 'sleep drive'. The longer you're awake, the greater

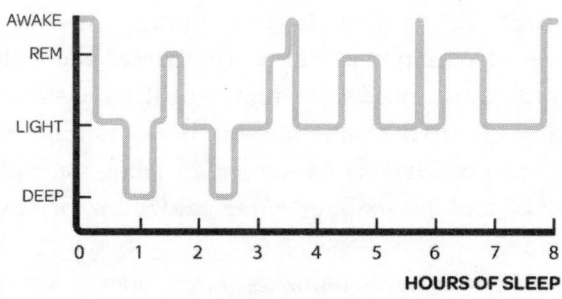

Normal sleep
© Dr Guy Meadows. Reprinted with kind permission.

the need or drive for sleep. So you need to be awake for around 16 to 17 hours, e.g. 7am to 11pm, to create enough drive to achieve 7 to 8 hours of sleep. It's easy to mess up the pattern, for instance falling asleep in front of the TV instead of going to bed, trying to catch up sleep by staying longer in bed in the morning or taking lengthy naps. Then when you do go to bed, you're often tired but can't sleep as the sleep drive hasn't built up to a sufficiently high level. Tired, frustrated, angry and worried – all these emotions trigger a feeling of being in danger and further feed wakefulness; they're the stuff that insomnia thrives upon. Likewise a 'good lie in' on a Sunday morning can weaken your sleep drive for Sunday night. Add to this Sunday night worries about not performing well and looking terrible at work on Monday, and you have the recipe for starting next week's insomnia.

UNDERSTANDING YOUR SLEEP WORRIES

It can be useful to determine how severe your insomnia is now, and then, when you've put the therapy into practice for about 6 weeks, you can rate yourself again to see what improvement you've made. The following questionnaire can also highlight the areas that are particularly troublesome for you. It covers the 5 steps developed by Dr Guy Meadows, a leading sleep physiologist, and his colleagues at the Sleep School in London.

Using the questionnaire, rate yourself on how true each statement is for you, with 0 meaning Never, 1 Sometimes, 2 Frequently and 3 Always.

Step 1 Discover

	0	1	2	3
I struggle to control the quality of my sleep.	0	1	2	3
I keep trying different things, like pills and rituals, but nothing works in the long term.	0	1	2	3
I don't understand why the things I use to help me get to sleep don't work.	0	1	2	3

Step 2 Accept

I dwell on past poor-quality sleep or worry about future sleep, and find it hard to switch off my racing mind at night.	0	1	2	3
I jump from one topic to another and struggle to maintain my focus and concentration in the day.	0	1	2	3
I just can't concentrate on how I am feeling, on my thoughts and on how my body is reacting.	0	1	2	3

Step 3 Welcome

I have negative thoughts about how poor sleep affects my life, and I struggle to get rid of them.	0	1	2	3
I experience strong emotions like anxiety and sensations such as knots in my tummy when trying to sleep, or the day after a bad night, and I can't control them.	0	1	2	3
I find it hard to resist unhelpful urges when trying to get to sleep, like watching TV, or taking a sleeping pill or alcohol.	0	1	2	3

Step 4 Build

I change my sleeping pattern to control my sleep, such as having a long wind-down, going to bed early or late, lying in or taking long naps during the day, sleeping alone or getting out of bed at night.	0	1	2	3
I need my bedroom environment to be perfect (e.g. light, temperature, noise/silence); otherwise I won't sleep.	0	1	2	3
I change my lifestyle (e.g. altering my diet or not socializing in the evening) in order to get myself to sleep.	0	1	2	3

Step 5 Live

I can't fully get on with my life until my insomnia has gone.	0	1	2	3
I feel that insomnia impacts my daily life negatively, affecting my relationships, energy, work and health.	0	1	2	3
I always worry that my insomnia will return and that I might have it forever, even when my sleep has improved.	0	1	2	3

Write down your total score and today's date ....

Now let's take a closer look at those worries. If you are reading this book in chapter order, you will now understand a lot about what anxiety is, where it comes from, the forms it takes, and what you can do about it (Chapter 3). While the first edition of this book described the more traditional CBT treatment for insomnia, recent research by Meadows and his colleagues suggests an exciting new angle. They've found fascinating parallels between the way we respond to anxiety and to insomnia, and as a result, propose basing treatment on ACT (Acceptance and Commitment Therapy), developed by Stephen Hayes, Kelly Wilson and Kirk Strosahl. This therapy, they suggest, is likely to be even more effective than the traditional CBT model.

Look at the diagram of 'How CBT understands anxiety' (page 18). Replace those 'Worries/anxious thoughts' with all your fears of the consequences of not sleeping that night – poor work performance, looking terrible, harming your health, no energy and so on. Pretty heavy stuff ! It's no wonder your body and mind responds to these as if they were imminent dangers. These thoughts then invite their friends – related thoughts, past memories and catastrophic predictions about the future. Your body's arousal starts climbing the scale, meaning considerable distress and discomfort, and you're in that vicious circle of worries feeding symptoms feeding worries. And, as you go round and round, sleep recedes ever further.

You can also see that the avoidance behaviours (getting out of bed to watch TV, have tea, take a bath, read, take pills) you've tried , probably unsuccessfully, have become part of the maintenance cycle of insomnia. They inadvertently feed your worries and your struggle to sleep, and your attempted cures become the things that are actually keeping you awake.

The research suggests that you have choices when it comes to treating insomnia and responding to what is going on in your mind and body. You can take new, different, even fun and playful options. Instead of the traditional guidance – finding 3 arguments against the thought telling you that you will come to harm, then trying to think about something else so as to leave behind those thoughts and emotions – ACT puts forward the idea that you engage with your worries. Have a look at Chapter 6 on Depression. In the section 'Thinking about thinking' (page 122), we describe how you can choose to buy into your thoughts, believe and even feed them, or how you can choose not even to argue, but rather to see them as thoughts, not facts, and use a mindfulness approach to distance yourself from them. Try out the exercise on pages 122–23.

Bet you didn't try it! Go on, do have a go.

Did that feel useful, and did it give you a bit of a break from your worries?

MINDFULNESS

Mindfulness is a key component of ACT, and is based on the idea that we need to concentrate more time and focus on the present. The present moment, where we are right now, is usually not an unpleasant place to be. However, minds often go into problem-solving mode, wanting to 'fix things' and trying thereby to prevent problems. Your brain often attempts to anticipate what could go wrong, in the hope of finding solutions in advance, or to dwell on the past and all the things that did go wrong, searching for ways to stop repeating this in the future. This can be very helpful for certain types of problems where you can learn from mistakes you made or can plan ahead. However, this mental activity takes a huge amount of energy, time and anxiety. It's important to be able to differentiate between those things that you can and cannot change, as in the Serenity Prayer:

God, grant me the serenity to accept the things I cannot change,
Courage to change the things I can,
And wisdom to know the difference.

If you're beating yourself up about past events you can't change, or anticipating catastrophes, most of which will never arise, then you're wasting all that energy and getting unnecessarily anxious. This applies to your thinking during both night and day. The most recent research suggests that good sleepers *do nothing* to sleep well – they just get on with their normal lives, and at their chosen bedtime, go to bed … and sleep! But insomnia sufferers frequently spend considerable mental activity thinking about their past and future sleeping difficulties, desperately seeing remedies to fix the problem.

Practicing mindfulness during both the day and evening lets you develop a detachment from the worries churning through your mind. There are many simple, effective mindfulness exercises for you to try out. Often one is more helpful than another – some people think in pictures, others largely in words – so have a go at several of the exercises below and see which best suit you. If you practice them during the day, away from the extra pressures of lying awake fighting to get to sleep – something you can now see is probably doomed to failure – you are likely to develop the skill of being mindful. You can then put the exercises into practice when you are lying in bed, becoming aware of your thinking and your environment, and just noticing them, as well as becoming aware of your senses, your breathing and the feelings in your body. You can then lie quietly at rest, observing yourself and feeling detached from what your body and mind are doing. Mindfulness is not about trying to get you to sleep. Instead, what it does is create a different mental state, one in which sleep becomes more likely.

Before we go through some of the exercises, be aware that your mind may try to interrupt you. Perhaps it will try to belittle the exercises ('this is all rather silly') or tell you that you're no good at it and it won't work for you. Or perhaps your mind will try to make you think about what's going to happen tomorrow ('that'll probably go wrong') or what didn't work yesterday. Whatever interruptions your mind comes up with, see if you can notice the thought, tell yourself that it's an interesting thought, and then slowly bring your mind back to what you are focussing upon. When it wanders off again, just repeat the process and keep bringing your mind back to your point of focus.

Here are a few mindful exercises. Have a go during the day; even 3 minutes in the morning and 3 in the evening can start to make a difference. Some can also be practiced when you go to bed at night. You may be surprised by how much fun they can be.

MINDFUL LISTENING

- Find a private spot where you won't be disturbed – it need only be for 5 to 10 minutes.

- Now close your eyes and concentrate on the sounds around you that are loudest.

- Listen harder and discover quieter sounds further away. Really pay attention.

- Now start using your imagination. Try to imagine you are hearing sounds like the rumble of public transport, conversations murmured in the distance, birds and beasts in faraway places, even other continents – anything at all you want to listen to, even music.

- When your allotted time is up – perhaps having set an alarm for those few minutes – gently return to the present and off you go with your daily life!

MINDFUL LOOKING

- This probably isn't one to try out in bed in a dark room!

- Go back to that private spot, or find a new one. It just needs to be somewhere no-one will bother you for a short while.

- Look around you. What catches your eye? Looking at whatever object or creature you have spotted, really focus on the colours, shape, size and shadows – everything about it.

- Then start to scan the rest of your environment, paying close attention to everything, one bit at a time. Try to create a photographic memory of it, accurate in all the details.

- Time's up! Back to your busy life – but hold on to the peaceful feelings.

MINDFUL LIVING

Whatever it is you are doing, really notice everything about the activity. Whether it's walking, sitting, eating or lying down, focus on all the physical sensations, sights, sounds, the temperature, in as much detail as you possibly can. You will probably find that things feel so much more real and vibrant when you do this.

MINDFUL RESTING

- This is particularly useful for those previously long, sleepless nights.

- Start by reminding yourself (again!) that *mindfulness is not about trying to get to sleep*. You are telling your body and mind that you are willing to experience the discomfort associated with not sleeping. If sleep creeps up on you, that's fine, but the aim here is to explore the mindful resting experience.

- Comment to yourself on your physical sensations – for instance, 'I can feel my heart beating really fast. There are butterflies in my tummy'. By doing this, you're accepting what your body is doing, without either judging it ('this is bad for me') or trying to change it (e.g. with relaxation exercises or medication).

- Notice your feelings and your thoughts. Just observing them enables you to detach yourself from the struggle of forcing sleep to come.

Remember that the harder you chase sleep, the faster it runs. If you fight it, it fights back! And you are the loser, because your body releases adrenaline to make your reactions stronger, meaning you become even more alert and aroused, and less likely to fall asleep. By choosing to be still and gently notice whatever arises, you're telling your brain that sleeplessness is no longer a threat. You're also saving loads of valuable energy, which will make an enormous difference to how you feel tomorrow.

Mindful resting can sound pretty daunting. It takes courage to put down your weapons – your 'bag of tricks' that you believed you required – and to face everything while you lie in bed, feeling scared and defenceless. But just like those anxiety symptoms in the last chapter, when you do face your fears, they so often turn out to be not nearly as terrible as you had anticipated. Remember that client who faced their fears and found out the ferocious lion was really only a pussycat? Now let's look in more detail at those pesky thoughts.

PUTTING ACT INTO PRACTICE

Your first step is about **acceptance**. Your mind is coming up with certain thoughts, and, try as you might, you cannot stop, let alone prevent this. Of course you'd like to be rid of them, but they're refusing to leave you alone. So, ACT asks you to accept them. Most of us don't want the unwelcome, upsetting thoughts. And when therapists talk about 'acceptance', it can easily be confused with resigning yourself to never being free of the thoughts, and with no hope of the future being any different.

But this is certainly not what ACT proposes. If you experience an unwanted thought, whether during the day or when lying in bed at night, the suggestion is that you notice it, using your newfound mindfulness, and then actually *welcome* it. Acceptance involves choosing to do nothing to try to get rid of thoughts. It's a positive act, whereby you purposely choose to do nothing because this is the most helpful course of action and the best way to solve the problem at this particular time. Rather than trying to stop thinking the thoughts, you might start to label the different types of thoughts – 'WW' are work worries, 'HA' are health anxieties, 'AP' are worries about your appearance, 'OP' are about other people ... and so forth. So, you may think something like, 'What's this spot on my arm? Skin cancer? That's a HA thought!' Or perhaps, 'That person was horrible to me today. They don't like me. Bet nobody really likes me ... Hey, there's a new OP I haven't had before! Welcome to the club!'

Playing with your thoughts allows you to detach and distance yourself from them, rather than buying into them and taking them as truths. It means that you can watch your mind with interest, particularly when you are lying in bed at night, and wait to see the latest and most original thoughts your mind can produce, or even welcome old friends – 'That's 2 APs, 1 HA, and no WWs ... so far!'

Another playful way of managing thoughts is to notice which thought you are having most strongly – perhaps it is 'there's no sleep for me'. You then start to sing the thought to a familiar tune, like 'Happy Birthday', or even the national anthem. This can make you feel very differently towards the thought, and can even make you smile. Not something you would have ever expected.

SUMMARY

We're halfway through the chapter, so it feels like a good time to summarize what we've covered so far ...

- Do nothing. Stop struggling against your sleep issues.

- Let go of trying to sleep. Lying peacefully is just fine, and far preferable to battling.

- Watch and observe your thoughts and feelings without judgement.

- Become more mindful and live in the now.

- Welcome and play with your worries and fears.

... and look at what's still to come.

- Decide on your new sleeping pattern.

- Stay in bed all night, enjoying the benefits of resting at night.

- Become a normal sleeper.

- Keep your good sleep on track and don't let set-backs derail it.

YOUR SLEEP PATTERN: CHOOSING WHAT SUITS
YOU

Has there ever been a period in your life when you slept well and felt refreshed on waking? If so, try to answer the following questions about that period:

- How many hours a night did you average then? (The norm is between 4 and 10 hours.)

- What time did you tend to go to bed?

- When did you generally get up, and did you get up with or without an alarm?

- What sorts of activities did you do (or even decide not to do) just before bedtime?

If you cannot remember such a time, think of close family – do you think any of their patterns are likely to suit you?

It's important to remember that a normal 'good night's sleep' does *not* consist of a solid unbroken 8 hours – for anyone. We all take a while to fall asleep, and all of us waken one or more times at night between those +/– 90 minute sleep cycles. Often the waking is for such a brief period that you hardly notice it and have forgotten it by morning.

When you have chosen what you think is likely to be your ideal sleep duration, shorten it slightly, perhaps staying up an extra half hour. This gentle sleep restriction can mean you spend slightly less time lying in bed awake, and so sleep more deeply and have more energy the following day. Keep to this pattern for a few weeks, varying it by not more than half an hour where possible.

Of course, exceptions will arise like celebrations or deadlines, but try to keep to your chosen, regular pattern. Then, after a few weeks, show yourself you can be flexible by either adding to or restricting the time chosen. Repeat this cycle of a regular pattern followed by additions or restrictions, every few weeks.

Stick to it! Show yourself you are willing to commit to the new you that you want to be, despite the likely initial discomfort. If you find your resolve weakening, remind yourself that sleeping late, or going to bed and getting up at irregular times resets your body clock to a new time, and so upsets the cycle for the next night.

PRE-SLEEP PHASE

At a personal level, the first author of this book will never forget the period of her life, aged around 12, when she was away from home for the first time. Every evening around 5pm she would look at the sun, praying that it wouldn't go down. She 'knew' that when the sun went down, night followed, which meant bedtime, and that she 'wouldn't sleep'. And sure enough – she usually hardly slept at all. The sadness of begging the sun not to go down every night is still vivid to this day, though fortunately the insomnia belongs to the past.

Your new pattern, which follows normal sleepers, will be about coming home, having supper, and then making time for leisure or some evening work. As bedtime approaches, you stop the more stimulating activities like watching TV, having family time and doing stuff on the computer. You start winding down, getting undressed, brushing your teeth, and maybe reading in bed or chatting to your partner. Then you switch off the lights and lie quietly, letting your mind wander at will. You'll find that you drift in and out of consciousness in this presleep phase, sometimes experiencing a muscle spasm or jerk, or even a burst of adrenaline that can feel like a bolt of electricity. These are all normal sensations many people have when waiting to fall asleep. You also might have the sensation of falling. This too is quite normal. If you're not alarmed by this, then your body won't respond as if to danger. You can just allow yourself to mindfully notice and label all the sensations and thoughts you are experiencing while in your warm cosy bed.

The most important thing to remember from this whole chapter about dealing with insomnia is this: *normal sleepers are content to allow themselves to be quietly wakeful in the pre-sleep phase.* They aren't trying to will themselves to sleep; they are confident sleep will come when it is ready. They also know that even if sleep doesn't show up, a peaceful night resting in bed while letting their mind wander is the next best thing, and can provide rest and be refreshing. It can even be relaxing! Paradoxically, the key to sleeping is having an accepting and relaxed attitude to being awake at night.

You have probably spotted the parallel with the way to handle panic attacks. You actually go looking for them! You seek out the biggest, scariest ones you used to believe would kill you, and find that:

1. They can indeed be unpleasant, but aren't harmful, and

2. The more you seek them out, the fewer you actually find.

One patient said at her follow-up therapy session: 'I'm so sorry. I looked absolutely everywhere in all the places I used to get panic attacks – and I couldn't find a single one!'

You may well find that when you practice your mindfulness, acceptance and welcoming for the whole night, sleep could sneak up on you and overtake you!

We understand that at the start of your journey to better sleep, it can feel too much to demand that you stay in bed all night long. It's helpful to allow some flexibility. Try sitting up at the edge of the bed, practicing mindfulness, for a pre-determined length of time, which you gradually shorten over time. Of course, do give yourself permission to go to the toilet, but notice if the frequency is excessive, as toilet visits can be avoidance in disguise.

The most workable way of building a helpful relationship with your sleep is by learning how to be in bed with your fears, rather than out of bed escaping them.

Dr Guy Meadows

BED SHARING

The ACT view on this is that sleeping separately from your partner, while perhaps improving your sleep in the short term, can have long-term consequences for the relationship. And starting this therapy in separate beds, waiting until you're confident that your sleep pattern is sufficiently secure for you to share a bed, might mean you'll never get that confidence, as sleep patterns are known to vary even for normal sleepers.

If the prospect of sharing a bed for a whole night is too daunting, start with a fixed time, like 30 minutes to an hour, and build up from there. Use your mindfulness to enable you to notice all the different sensations and experiences of sharing a bed, and be prepared to welcome each as you notice them.

IT'S MORNING!
It may sound obvious to advise a wake-up routine – lights on, curtains open, getting washed and dressed and having breakfast before heading for work, school or the day's planned activities. But many insomniacs are so desperate to grab any sleep at all that they will stay in bed until the last possible moment, then head off to wherever in a chaotic rush. Not the most conducive start to a peaceful and successful day, would you agree?

If you haven't slept much, of course you will be tired during the day. Research suggests a 10- to 20-minute nap, after lunch but before 3pm, does no harm and can even do you good. Mindful waking in a quiet place is equally effective. For those who are ill, elderly or pregnant, a nap of up to 90 minutes (in other words a whole sleep cycle) can make a big difference to their whole day, and night. You will need to experiment to find what works best for you. For many people, gentle exercise is a good idea when you feel really tired. A short 20-minute (mindful) walk in the fresh air can make an enormous difference to how you feel, and how you sleep that night.

LET'S TALK ABOUT LIVING
If you choose to live your life to the full, the way you wish to do, despite suffering from insomnia, you take away all the reasons for struggling with sleeplessness.

Less struggle means less night-time arousal, less wasted energy and therefore more capacity for living a rich and meaningful life. This life then creates an environment from which natural, health-promoting good sleep can emerge.

Giving up part or most of your life in order to control insomnia, so that you can go back to living your life, just isn't a workable equation, but rather another vicious circle. The paradox is that if you are not prepared to accept the discomfort of your sleeplessness, you may never experience good quality sleep. A life without pain and discomfort comes at the cost of life itself.

Think about your life right now. What are you not doing that you would be doing, if you didn't have insomnia? Here are some things others have said:

- Share a bed with my partner

- Play more often with my children

- Go out more, meet new people and perhaps find a partner

- Start that fitness regime

- Socialize more with my friends

- Feel confident to get pregnant and start a family

- Work hard and try for a promotion

- Start that hobby I've wanted to try for ages

- Sign up for an evening class or two

- Get down to some serious studying to pass my exams.

What's stopping *you*?

ACT is a way of moving your life towards things that are valuable to you. Look at the list above, or choose something else, and in the next 24 hours *do it*!

RELAPSE MANAGEMENT

Overcoming your problem does not vaccinate you against further bouts in the future. So, while you are not immune, neither is a normal sleeper. In fact, it's normal for most of us to have occasional periods of difficulty sleeping. But this need be no worse than having a bad cold or flu. You know what it is, how to manage it, that it won't harm you and need not mess up your life. So you are able to deal with it, knowing that eventually it will pass, while you build and live the life you choose, irrespective of whether insomnia has dropped by for a visit.

Just before you complete this chapter, after you've spent a few weeks putting its lessons into practice, go back to that questionnaire on pages 59–60. When did you fill it in? You may find it helpful to get a notebook and record your results monthly for the next 6 months. Hopefully you will impress yourself with the changes you've made, as well as identifying anything that looks like it needs a bit more work.

In conclusion, you now know how to mindfully accept what your mind wants to do, to welcome whatever it comes up with, and to be able to live a full life of both happy days and peaceful nights – and peaceful days and happy nights too!

So, go on! Kiss insomnia goodnight – and sleep tight!

CHAPTER 5

Beating bad habits and building better ones

The unfortunate thing about this world is that good habits are so much easier to give up than bad ones.
Somerset Maugham

Practice, it is said, makes perfect. When we deliberately keep repeating an activity, it may be in order to improve our skill and get to the point where we can do it with minimal thought. Habits are also things we do regularly, often without even thinking about them. This ability to do things almost on autopilot, with minimal conscious thought or control, is very important to the way we live our lives.

Consider learned skills – imagine trying to walk up a flight of stairs, tie your shoelaces or drive a car if you had to think through and plan each stage of the process every time you did it. Learning a new, complex skill like driving is really tough at first, requiring serious concentration. Every stage has to be thought about and you can feel overwhelmed by how many things you have to do at once. However, once you have developed the skill, you no longer need to think about each step – you can carry them out automatically, freeing you to concentrate on other things such as where you are going or avoiding the neighbour's cat that runs into the road!

Likewise, being able to develop habits is very useful. For example, if you consciously practice putting your clothes in the wash-basket when you undress, rather than on the floor, and deliberately ensure you put an object back where you found it after using it, you are well on the way to developing the habit of tidiness. However, the downside to this is that we can equally develop some habits which aren't so useful. There are many different types of 'bad' habit, which may be bad in different ways. Some relate to how we treat our bodies – such as using addictive substances which are bad for us, eating the wrong foods, not drinking enough water or taking too little exercise. We might bite our nails, pull our hair out or pick at our skin. Others may be interpersonal habits. Think for a moment about how you treat other people. Are you highly critical? Do you irritate your partner by being thoughtless about particular things around the house? Maybe you are always late or untidy. All of these involve patterns of behaviour which can become habitual. The bad news is that it takes effort to change those patterns. The good news is that it can be done, often with great success.

Now let's get personal! Think about your bad habits. Write them down in the following table.

Bad habits I've had up to now and would like to change

A WORD ABOUT ADDICTION

Habits that involve addictive substances such as nicotine (found in all forms of tobacco), alcohol and other drugs can be overcome using many of the techniques we discuss in this chapter, which are effective for changing all habits. However, substance misuse can be more than a habit. It can lead to addiction and if you are addicted to a substance you may need additional help to give up, besides learning how to change your habit.

Take smoking. There are various alternative ways of getting nicotine other than via tobacco. Many people find it helpful to stop their tobacco habit by initially still having nicotine, for example in chewing gum or patches. They then gradually reduce their intake of nicotine, withdrawing slowly from the drug, having first broken the habit of smoking, or any other form of tobacco use. If you do want to give up tobacco, in the first instance perhaps you might try out the options we provide and not take nicotine in another form, to test out whether you actually need to slowly withdraw from the nicotine. The actual effects of nicotine addiction are usually relatively short-lived and the physical cravings pass within a few days – it is then the habit itself of using tobacco which needs to be broken.

Be aware that drug use of any sort can be quite difficult to change in the initial stages. You need to break both the addiction and the habit at the same time – and you are having to do this at the point where you are least confident in your ability to succeed. Withdrawal from addiction to alcohol or other drugs can be so difficult that some people need professional help managing the process. Consider seeing your doctor or an addictions specialist. There are many sources of help out there (see Chapter 10). So, if you find it impossible to give up a substance by using only the techniques described here, do go and ask for help. Once your body overcomes the physical addiction to a substance you can then concentrate on breaking the habits, both behavioural and cognitive (what you do and what you think), which you have built up around your addiction.

HOW MUCH IS TOO MUCH? HOW CAN I TELL IF I'M DRINKING TOO MUCH?

Many people actually know that alcohol abuse is a problem for them, but have difficulty admitting it to themselves or others. Keeping an eye on whether you are regularly drinking more than the recommended number of units for your gender is important, but so is examining your drinking behaviour. It is estimated that 1 in 25 adults in the UK is alcohol-dependent.

Answer the following questions. Do be honest – it's private to you, and can highlight whether you might need to do something about your drinking:

1. Do you find it very hard to stop drinking?

2. Do you get physical reactions when you abstain from alcohol?

3. Has anyone around expressed concern about your drinking?

4. Do you always drink when stressed?

5. Do you avoid events or situations where you know you won't be able to have a drink?

6. Do you regularly do things whilst drinking that you regret when sober?

7. Do you find it hard to go for 'just a half', or stick to one glass of wine?

8. Do you almost prefer to drink alone?

9. Do you ever feel you absolutely must have a drink to get through certain situations?

10. Do you or anyone else have to actively set limits on your drinking?

11. Do you regularly drink in order to get rid of a hangover?

If you have answered 'yes' to any of these questions then you may have a problem with alcohol. If you think you do fall into this category then we recommend you get advice from your doctor.

STEP 1: DECIDING TO CHANGE

Habits are difficult to change. As we have seen there is a good reason for this. They need to be firmly established and quite resilient to enable us to do things automatically. However, this means that though we may make a change, it's oh so easy to slip back into our old ways if we are not on our guard. In order to make a difficult change, and maintain it, we need to be motivated to do so – if we are not then we are never going to be able to undertake the hard work required.

WEIGHING UP THE COSTS AND BENEFITS OF HABIT CHANGE
Taking some time to really think about and explore your motivation for changing a habit will make it much more likely that you will succeed. Think through all the advantages and disadvantages of changing. This table can help:

	Benefits (pros)	Costs (cons)
Changing	☺	
Staying the same		☺

Filling in each section of this table can help you to think through your reasons for changing. Make sure you work out the benefits (pros) and costs (cons) both of staying the same and of making a change in your habit – these may often be direct opposites of each other but aren't always. If you miss out a section you might miss some important factor in helping you to change. The smiley faces are in those sections which may contain the factors most likely to motivate you to change. If there are lots of strong benefits to changing, and high costs to staying the same, then you are more likely to be able to make this change. Do some work on this – ask other people around you what they think as they may identify some areas which you haven't considered.

Of course whether you feel motivated to change won't just depend on having lots of things written in the 'benefits of changing' and 'costs of staying the same' boxes. The *weight* of the individual things will count too. You might only have one benefit to staying the same, but if this is something that is hugely important for you then changing will be hard no matter how many things you have in the other boxes. Think about the weight of each identified pro and con. Are there ways in which you could change them? What information do you need to gather to help increase the weight of the things that might help you to change this habit?

Here's an example of how this might work.

case study ANNA-MARIA

Anna-Maria is a smoker, aged 31. She started smoking at school and has continued for many years. She generally smokes around 20 cigarettes a day. She actually enjoys smoking and believes that it relaxes her. Many of her friends smoke, although these days more and more are giving up. Anna-Maria knows smoking is bad for her. This was brought home very strongly when a family member who had been a life-long smoker was recently diagnosed with lung cancer. Anna-Maria wants to stop smoking and has tried before but found it too hard. When she tried previously she found that, though the cravings passed quite quickly, it was actually the habit of smoking which seemed impossible to give up. Anna-Maria is married with two young children. Let's look at how Anna-Maria used the pros and cons exercise:

	Benefits (pros)	Costs (cons)
Changing	☺ • Improved health • Less breathless – might be able to start exercising more • Won't smell of smoke • Will save money • Children and husband really want me to stop	• Have to deal with cravings • Will miss the social side of cigarette breaks at work • Find smoking relaxing – might get more stressed • It's really hard to change
Staying the same	• Needn't go through all the effort of giving up • Enjoy time at work with friends who smoke	☺ • Passive smoking could affect the children's health in future • The children might copy me and think it is okay to smoke • Husband gets angry with me for spending money we don't have • Health might suffer

Anna-Maria works through this exercise. She notices that there are several things that are very motivating in both the benefits of changing and the costs of staying the same. The most important are those relating to her children. She finds it helpful to look up information on the dangers to children of parents smoking and on the health risks of cigarettes. This information greatly adds to the weight of those factors in motivating her to change.

She also notices, however, that there are some quite powerful things in the boxes on benefits of staying the same and the costs of changing. In particular she believes strongly that smoking relaxes her and that she'd become very stressed if she stopped smoking. She looks more deeply into this, and examines the evidence that this is true. Her reading helps her to question this assumption.

She realizes that it might be helpful for her to develop other ways of dealing with stress that are better for her – she takes up yoga, finding that she really enjoys it. She talks to her husband about his role in helping her with stressful events at home. She also tackles her worries about the social side of smoking. She discusses how to spend time with her friends without joining them for a cigarette break. She also persuades one of her friends that it might be a good idea for both of them to give up together – she finds this really boosts her motivation.

REMEMBER
Unless you really want to change you will not be able to. It's worth putting in some work at this stage in order to make change possible.

EXAMINE YOUR THOUGHTS ABOUT CHANGE
Throughout this book we emphasize the importance of thoughts and how they affect the way we feel and behave. Obviously habit change involves changing behaviours, and for this we suggest a number of behavioural techniques, many of which are not exclusive to CBT. However, often what trip people up when they are trying to change old habits or create new ones are the negative thoughts they have about change, or about their ability to change. Frequently they're quite unaware of these thoughts. Spend some time thinking about what thoughts, beliefs or assumptions might be holding you back from making the habit change you want.

case study DEVIKA

Devika wants to lose weight. She is several stone overweight and this is affecting her health and self-esteem. Her relationships suffer because she feels so bad about herself. She avoids both social contact, and intimacy with her partner. Her breathing is affected and she has developed diabetes. Most of the time she is very motivated to change. She makes a good start, eating more fruit and vegetables and going swimming twice a week. However, she still often finds herself eating junk food like crisps or chocolate bars when she is in a hurry or busy. She needs to change this habit to make more progress. Devika examines her thoughts about eating. She notices the following pattern:

> **Thought:** I'm hungry, I want crisps.
> *Triggers …*
> **Thought:** You shouldn't do that. You are a pig. You are fat and disgusting and lazy. *She gets a picture in her head of her mother criticizing her.*
> *Triggers …*
> **Emotion:** guilt, sadness, shame.
> *Triggers …*
> **Behaviour:** dwelling on negative thoughts and memories of the past.
> *Triggers …*
> **Thought:** It's not fair. I never get to have what I want. Everyone else can eat what they like. I never get what I want. *She pictures her mother scolding her for eating crisps.* I'll show them – I'll do it anyway.
> *Triggers …*
> **Emotion:** anger. *Triggers …*
> **Behaviour:** she eats the crisps.

Devika realizes this has become a vicious cycle of guilt, shame and anger which leads to her behaving in ways she doesn't actually wish to, and which results in her feeling worse. Once she's aware of this she's able to break the cycle. She recognizes the self-critical thoughts that make her feel guilty and challenges them. Instead of responding to these negative automatic thoughts she reminds herself that actually she's been working very hard at changing things, that her partner has always found her attractive, whatever her size and that she's successful in many areas of her life. She gathers evidence for herself to bolster these more positive thoughts.

Devika finds that challenging these self-critical thoughts means that they no longer trigger ruminations about things from the past which make her both angry and more likely to eat the wrong things. It's more helpful for her to be kind to herself and to focus her thoughts on the fact that she is now *choosing* to eat differently because *she* wishes to – not because others are bullying her into it. This means that she feels much more able to follow her diet and to avoid slipping into old patterns such as rebelling by eating crisps.

When we find it a struggle to do things that we know are right for us, it's usually because there are patterns of thoughts, emotions and behaviours like these holding us back. Identifying them and then working on changing them can enable us to make the alterations we'd like.

We may also have negative automatic thoughts about our ability to change. Anna-Maria in our first case study had thoughts that it was too difficult to give up smoking and that she was too vulnerable to stress to be able to do it. These thoughts held her back. She found it very helpful to do some preliminary work on testing out these beliefs before embarking on trying to give up smoking. Once she began to be convinced she had other ways to handle stress, she was more able to put in the energy needed to break her habit. Using the thought challenging and behavioural experiment sections from Chapter 3 and Chapter 6 would be helpful here.

STEP 2: KNOW YOUR ENEMY

Habits are automatic – that means we do them without thinking about them and are very often unaware of why we do them or what sets them off. Take biting your nails for example – many people do this when they are anxious or stressed, others do it when they are bored or distracted.

UNDERSTANDING AND MONITORING YOUR HABIT

In order to begin to change a habit, you need to identify the following:

- When do you do it?
- How often?
- What are the triggers?

BECOMING MORE AWARE

'Autopilot' is a mental setting that is very useful at times. However, it becomes the enemy when you are trying to change a habit. We can learn to switch from 'autopilot' to 'manual' by teaching ourselves to become more aware of our habits. This takes time and practice.

Key tips:
1. Remind yourself daily to look out for your habit.
2. Write a note to yourself and leave it somewhere you will see it regularly.

3. Ask people to remind you.

4. Note down anything else that helps increase your awareness.

KEEP A HABIT DIARY

One helpful way of increasing your awareness is to **keep a diary of your habit**. Each day monitor when you indulge in your habit and what is happening at the time. Look out for the *urge* to indulge in the habit, as well as noting when you actually carry it out. Can you recognize any triggers for it – what were you doing, thinking or feeling at the time? All of this information will help you to build up a picture of how your habit happens and the times or situations that you need to be particularly wary of when planning your habit change. You can include anything in your diary that looks like it could be useful information. The important thing is to note down whenever you catch yourself indulging in your habit. It may take some practice and discipline since, as we know, some habits may happen without our awareness. See how often you can catch yourself 'at it again!'

Self-monitoring of any kind quite often alters whatever we are trying to measure. For example, have you ever tried to keep a food diary or to write down everything you spend in a week? If so, you might have found that you change what you eat or spend because you become more aware of it just by keeping the diary. Bear this in mind when trying to understand more about when and how you indulge your habits.

case study MARIO

Mario sucks his thumb. He has done this since he was a small child and has never given it up. It used to soothe him when he was a baby and he still finds it relaxing. Mario is now in his teens and this habit often embarrasses him when people notice it. Sucking has made the skin on his thumb go dry and wrinkled, and sometimes it's quite painful. Mario worries that his hand looks ugly, and also that others will think he's childish. He starts a habit diary where he records whenever he notices he's sucking his thumb:

Day/time	Situation	Thoughts or Feelings	How long for?	What made me stop?
Mon 4pm	Sitting in a maths lesson	Really bored	About 10 mins	Ben laughed at me
Tues morning	On bus	Daydreaming about holiday	??	Noticed I was doing it
Tues 3pm	English lesson	Nervous about being asked to read	Few seconds	Had to pick up book
Thurs evening	Watching TV	Bored. Irritable. Cross with Mum for not letting me watch film.	About 20 mins	Mum told me to stop

Once Mario has continued this diary for about a week he can identify several patterns in his habit. He notices that he frequently sucks his thumb when he is bored or daydreaming and when he experiences unpleasant feelings like being irritated or anxious. He also notices that he doesn't suck his thumb when he's busy, occupied with something interesting or when he's in a particularly good mood. This monitoring helps Mario develop a plan for stopping his habit.

GET ANOTHER PERSPECTIVE

It can be helpful to include the views of people around you when you are trying to understand more about your habit. If you feel comfortable doing so, ask others what they have noticed about your habit and their thoughts about what triggers it. They may have noticed things you haven't! Their ideas can also be very helpful when you start actively trying to break your habit.

STEP 3: CHANGING YOUR HABIT

Okay, you are now ready to actually make your change. The following techniques will help you to do this. Be systematic in how you go about tackling your habit. Be persistent. You'll probably find making a written plan of how you are going to go about it will also be very helpful.

A – A SINGLE STEP AT A TIME

How many of us have set out on a 'health kick' or 'organization drive' where we try to completely change all of our behaviours at once? And is this approach ever successful? Very rarely. The reason that so many New Year's resolutions go awry is probably that people try to change too much. Just **concentrate on changing one thing at a time**. It takes effort and time to change even a small habit – try changing too many things at once and the chances are you won't succeed with any of them.

A single step at a time can also sometimes mean breaking a habit down into smaller parts and changing each of them, one by one. Consider changing things slowly – many people, for example, find it easier to cut down on smoking or eating certain foods before getting rid of the habit altogether. However you decide to do it – think it through and write a clear plan for yourself of when you will do what. It may be helpful to start small. For example, if you want to drink more water, start with one glass a day, at the same time each day. Keep this up for several days then add in one more, and so on. Don't try drinking eight glasses a day from the start.

B – BE REALISTIC

We probably could all do with being healthier, more patient, more organized and implementing many other such improvements. Everyone has habits that may not be ideal. Just as it's important to aim for changing just one thing at a time, so it's important to **aim for change that is realistic and achievable**. Don't try to completely reinvent yourself – the chances are you are mostly fine as you are – you just need to tweak the things you are not happy with. If you aim for 'good enough' in your habits you have every chance of succeeding. Aim for 'perfect', and you are setting yourself up for failure. It may be realistic to give up cigarettes. It may not be realistic to also never eat another chocolate bar, drink 3 litres of water every day, go to the gym five times a week and never shout at anyone.

C – CLARITY: BE CLEAR ABOUT WHAT YOU ARE CHANGING

Use what you've learnt from your monitoring to **make sure you are very clear with yourself** about what aspects of your behaviour you are changing and what your goals are. For some habits this will be obvious – you want to stop smoking or biting your nails or throwing your clothes on the floor. For others it may be less so – some interpersonal habits, for example, such as nagging, arguing or criticizing might take more clarifying. Be very specific about exactly what you want to be different and what you want to be doing or saying instead.

D – DATE: SET A DATE

One of the ways you can motivate yourself and get your energy levels up to tackle this habit change is to **build in some anticipation**. Set a date for when you will make your change and start anticipating it. Let other people know what you are planning – it can make it harder to back down then. Generate some excitement in yourself about this date. This is the day that you are going to start a process that will make you feel so much better in the long run, even if it's hard work to begin with.

E – ELEPHANTS NEVER FORGET ...

But you are not an elephant! Find ways to remind yourself why you are doing this. Remember your pros and cons list? Write down all those reasons why you are trying to change your habit and keep the list somewhere you can look at it regularly. Stick it up on the back of a door, the bathroom mirror, the fridge or your computer monitor – somewhere you will be reminded of it several times a day. **Write down your goal and keep that around as a reminder.**

F – FALLING OR FAILING: WATCH OUT FOR 'DANGER TIMES'

From your diary you will be able to work out your habit 'danger times'. For Mario, these were times when he was likely to be bored or have time to day-dream. **Be**

on your guard at these times. Avoid them if you can and if you can't, switch off your 'auto-pilot'– try to stay alert in order to catch yourself before your habit kicks in. Try to identify the urge to engage in the habit before you start. When you feel the urge, move on to something else.

Mario found keeping his hands occupied at times when he was day-dreaming or drifting off was helpful.

He held a stress ball while listening at school or sitting on the bus, and found that squeezing this worked as an alternative to thumb-sucking. Sometimes we can't predict when the trigger situations might happen. If your habit tends to occur when you are upset, you can't always see this coming. However, you can train yourself to be more aware that when you are upset you might engage in your habit and to be on guard against slipping into it at those times.

G – GOODIES!

Include in your plan small ways in which you will **reward yourself** after you achieve each milestone. This might be with something nice to eat, (not chocolate, if your goal is losing weight!) a small gift to yourself (unless your goal is spending less!), time spent watching a film or reading a book, a hot bath – anything that will feel like a treat and that you'll look forward to. It doesn't have to be anything spectacular but it should be an indulgence – not something you do every day, or would be doing anyway. Be proud of yourself, but watch out for self-criticism. Don't tell yourself your milestone is not an achievement because you should be doing the right thing anyway.

H – HELD BACK: WATCH OUT FOR NEGATIVE THOUGHTS

Remember the thoughts we identified which might be holding you back? Watch out for your mind ambushing you with these as you go along. These can be as automatic as the habit itself. Telling yourself, 'I'll never do this. It's too hard. I've always failed before.' won't help you to reach your goal. Raise your awareness of these thoughts. When you notice them starting to creep into your mind, let them go. Try instead to visualize yourself changing your habit. What will you look like when you have done it? Visualize the attractive, well-shaped nails you might have, or the way in which your partner will smile when noticing your perfume or aftershave, instead of the smell of smoke. Anna-Maria found it helpful to tell herself 'I am not a smoker' – she wrote it on the front of her work diary and found herself looking at it and repeating it whenever her habit threatened to drag her back in.

I – INVITE SUPPORT

Ask people around you (those whom you trust to be supportive) to help boost your efforts. Agree in advance with them what they will (and will not) do. Perhaps they can point something out to you when you don't notice you are doing it. Maybe they can help with rewards after an achievement. Watch out for problems here, though. There is no point in asking anyone to point out when you are picking at your skin if you are going to shout at them for doing so.

J – JUGGLE THINGS AROUND

Replace bad habits with better ones. Find something else to do with your hands instead of picking, scratching or pulling. Eat or drink something different. It can be much easier to replace a particular behaviour with another rather than simply doing nothing.

case study JUSTIN

Justin drinks too much. He is not addicted to alcohol but sometimes feels he may not be far off. He drinks almost every evening and has at times not been particularly alert at work first thing in the morning because of a 'heavy night' the night before. He works on a building site and occasionally his boss has raised concerns about safety when he appears hungover. His girlfriend has told him she worries he drinks too much, and his GP points out that he's drinking above the recommended limit for a man of his age.

Justin is also slightly overweight and, in trying to lose weight, he identifies that most of his excess calories come from drinking alcohol. Justin decides he wants to cut down on his drinking and works through the pros and cons exercise with his girlfriend to help him make a change. He recognizes that he is afraid of losing some of his social life and certain friends if he gives up drinking. He believes some friends would laugh or make fun of him for giving up. He examines his thoughts about this, and then tests them out by talking to some of his friends.

Most are actually far more supportive than he'd expected. Only one is disparaging and Justin realizes he cares less about this friend because of it. One friend even offers to actively help Justin by doing other things socially rather than going to the pub.

Justin keeps a diary of his drinking. He identifies that he drinks more when he is under stress or when he is with particular people. He decides it will be helpful to find other ways of relieving stress, and also arranges to see those people only with their partners present, since this helps him limit what he'll drink.

Justin sets a date to cut down on his drinking and informs close friends about his plan. He also decides on his target intake of units per week. He and his girlfriend identify outings and small rewards for each time he reaches a pre-agreed milestone. He avoids his 'danger times' and goes out running more often, as he has previously identified that this helps him to feel less stressed.

After a few weeks, Justin has reduced his drinking to a much more sensible, manageable level. He still enjoys going out with the boys from work, but does so less, and discovers when he's out with them he doesn't want to drink the quantities he used to. He has established a new habit.

DEALING WITH LAPSES OR SLIP-UPS

Whether we are breaking old habits or building new ones, we rarely get it right straight away. Lapses are inevitable and it is important that you recognize this from the beginning. When you have a lapse it can be tempting to be furious with yourself or to feel totally despairing and think you are just not up to the task. Both reactions can lead to you wanting to just give it all up. Try not to do this. You are only back to square one if you choose to be. Having a lapse doesn't negate the learning and achievement that went before – all of that still happened. Remember what you have achieved so far, even if it seems very little, and start again. Think about what caused the lapse. Identify your thoughts and feelings about it and learn from them. How can you incorporate this knowledge into your plan to make it less likely that you will slip-up again? Lapses are only a problem if you don't learn from them.

CHAPTER 6

Dealing with depression

*Depression is a prison where you are both the
suffering prisoner and the cruel jailer.*

Dorothy Rowe

'Depressed' is a word which we often use in everyday language to describe our mood when we feel low, out of sorts or just generally not at our best. For those who experience true depression, however, it's very different indeed. If you are depressed, the way in which you view and experience the world can change beyond recognition. Simple tasks become like climbing Everest, and the smallest of setbacks can feel like the end of the world. People have described the experience of being depressed as 'like wading through treacle' or like seeing the world through dark glasses, where colours become shades of grey. Winston Churchill described his depression as being like an enormous black dog which followed him everywhere, weighing him down.

Everyone feels low from time to time – it's completely normal. Arguably part of what makes us human is the way in which we experience a range of emotions in response to life's events. To feel sad in response to losses or disappointments or simply just to have a 'bad day' is something which we all experience, and usually just accept. Generally, when we feel this way we are able to identify roughly why we feel as we do. Usually the feeling passes reasonably quickly, or we do something to cheer ourselves up. It's important not to label normal sadness as pathological in some way.

However, for some people, low mood persists for weeks, months or even years with very little relief. It may be accompanied by other distressing symptoms, such as changes in appetite and sleep patterns, fatigue, physical aches and pains or feelings of worthlessness, helplessness and alienation from those around. Studies suggest that 1 in 5 of us will experience depression at some point in our lives. The experience of depression often brings other difficult emotions like guilt, shame or anger with it.

SYMPTOMS OF DEPRESSION
Do you recognize any of these?

- Prolonged depressed mood
- Thoughts of hopelessness and worthlessness
- Changes in appetite – eating too much or too little
- Changes in sleep pattern – sleeping too much or too little
- Tiredness and fatigue
- Physical aches and pains
- Restlessness or agitation
- Feeling physically slowed down
- Thoughts of self-harm or even suicide
- Difficulties with concentration, memory and attention.

It's a pretty gloomy picture. However, there's some good news in all this gloom. We now recognize and understand depression far better than just a few years ago. The stigma surrounding this and other mental health problems is slowly reducing. Sufferers often feel more able to talk about what they have experienced and to share their trials and their coping strategies. A recent campaign, 'Time to Change', highlighted the importance of challenging the stigma surrounding mental health problems like depression. It drew on the stories of famous people such as Stephen Fry, Alistair Campbell and Ruby Wax, demonstrating that **depression can and does happen to anyone**. But the very best news of all is that there are now many tried and tested treatments which have helped millions of people throughout the world. These include medication, talking therapies of many kinds, complementary or alternative therapies and community or social activities. CBT is one of the ways in which people with depression can be helped to get better, and evidence shows that it's one of the most effective.

WHEN DOES 'NORMAL' LOW MOOD BECOME DEPRESSION?

If you experience a difficult life event, such as a bereavement or other loss, such as redundancy, it would be unusual not to experience some low mood. These feelings may last some time, making it difficult to judge what might be considered 'normal' for such a situation. Often what you need most is time, support and care from those close to you. Most of us recover over time from such difficulties. However, for some people events like these can trigger a more lasting and pervasive depression. Usually you are considered depressed and perhaps needing help if you have had psychological and physical symptoms of depression on most days, over several weeks.

The following questions might help you think about whether or not you may be experiencing depression. Think about how you have felt over the past 2 weeks. Have you regularly experienced any of the following:

- Feeling very sad or irritable?

- A loss of interest in things that you previously enjoyed?

- Feelings of guilt or feeling bad about yourself?

- Being unable to concentrate, remember things or make decisions?

- Changes in your weight or appetite?

- Changes in your sleep pattern?

- Fatigue or feeling drained of energy?

- A feeling of restlessness or decreased activity which has been noticed by other people?

- Feeling that your situation is hopeless or that you are worthless?

- Thoughts of death or suicide?

If you have been experiencing more than 5 of these symptoms on most days over the past 2 weeks then you may well be depressed. If you think you are depressed it is a very good idea to talk to your doctor and explore the options for treatment. Self-help using the ideas suggested in this book can be an important part of treatment but should be undertaken in conjunction with professional help.

SUICIDAL THOUGHTS OR IDEAS

It can be very frightening if you or someone close to you experiences thoughts of suicide or self-harm. We don't often talk about these things but they are much more common than you might think. During bad times, huge numbers of us may have thoughts like wishing we could just go to sleep and never wake up, or that we'd never been born. There is a huge difference between having these kinds of thoughts and actually devising a plan to harm yourself. If your thoughts start to turn to plans, it's a good indication that professional help is needed. In a crisis, your GP or hospital A&E department are good starting points to find help. Research shows that depression is linked to chemical changes in the brain, making it very difficult for you to think in a clear, positive or healthy way. It is so important that we all get help and support at times of crisis, and don't act on depressed thinking.

HOW CBT CAN HELP US TO UNDERSTAND LOW MOOD AND DEPRESSION

If you've been depressed, you'll probably recognize how the 'five areas' model of CBT can help you to understand what's happening to you. Here's an example of this:

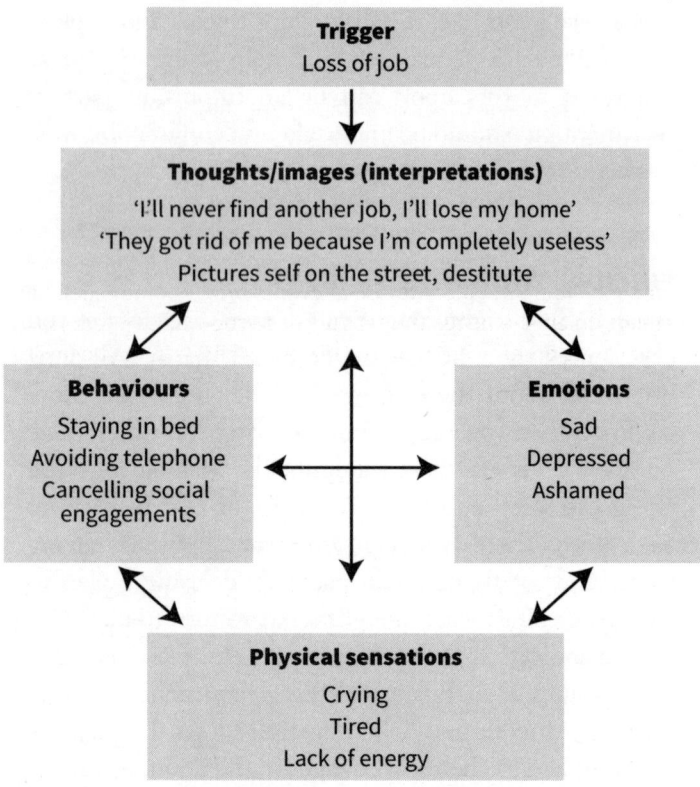

The 'five areas' model of CBT

case study LINDA

Linda is made redundant. The company is going through troubled times and Linda was the last to join so it's easy for them to let her go. However, Linda has always experienced low self-esteem and worries about her competency at work. This dates back to experiences of being bullied at school. Losing her job brought back feelings of not being good enough, and she tells herself that the company's been looking for any excuse to get rid of her. She feels ashamed, choosing not to talk to friends and family about it because she's sure they'll agree with her thoughts, and judge her. This means she's not hearing other points of view about her problems and finds it impossible to see any other interpretations of what's happening. Linda feels exhausted and drained of energy, stops going out and spends a lot of time in bed or alone, dwelling on negative thoughts. At first, avoiding people and social events brings some relief. She tells herself she needs the rest because she was so tired. However, over time, rather than the tiredness lifting, Linda begins to feel less and less motivated, ever more lethargic and weary, and her mood worsens.

WHAT CAUSES DEPRESSION?

There is no single cause for depression. People talk about depression as a 'chemical imbalance' in the brain and research shows that certain brain chemicals change in people in a depressed mood. However, for most people the cause of depression is a complex mix of biological (chemical), psychological (thinking and feeling) and social (life) factors.

CBT suggests that certain early experiences and life events (such as trauma in childhood, early losses, bereavement or bullying) can make us more vulnerable to developing depression later. This is because such experiences lead us to develop underlying negative beliefs about ourselves, other people and the way the world works. Let's look at how Linda's depression may have developed.

EARLY EXPERIENCES:
Bullied at school. Laughed at by other children.
Not well supported by parents – told she was being 'weak' and should just 'ignore it'.

NEGATIVE BELIEFS:
I'm weak and can't look after myself.
I'm incompetent.
Other people are likely to hurt me.
The world is unfair.

STRATEGIES FOR MANAGING THE NEGATIVE BELIEFS:
She avoids getting close to people. She sets really high standards for herself and works harder to avoid people realizing that she is incompetent.

TRIGGER FOR NEGATIVE BELIEFS:
Losing her job. Linda's underlying negative beliefs developed when she was quite young. However, up until now she's coped by employing strategies to prevent what she viewed as her incompetence showing to other people. The problem came when losing her job triggered these negative beliefs and led to a series of negative interpretations of her situation. This then led into the vicious cycle of thoughts, emotions, physical feelings and behaviours which we looked at above.

HOW CBT CAN HELP YOU TO TACKLE DEPRESSION OR LOW MOOD

The following ideas can help you improve your mood, whether you are suffering from long-term depression or just having a bad day. It is really important, though, to remember to ask for help if things are too much to manage alone. There is excellent help out there. Chapter 10 will help you discover where to get this help.

REMEMBER

Depression is incredibly common and sufferers are not alone. Ask for help if your depression is severe, if you are finding it impossible to begin to feel better or you have thoughts of ending your life.

STEP 1: GET ACTIVE!

try it now MOOD AND MOVEMENT

How would you rate your mood right now? On a scale of 0–10, where 0 is not low at all and 10 is the most depressed you have ever felt. Okay, write that down. When we feel very low, we don't want to do anything at all. Often we find it hard to muster the energy to even get out of bed. However, if you've scored 1 or more, here's the experiment: go out for a 5-minute walk, right now. Yes, even if it's raining! It's only 5 minutes, so you won't melt ... or drown!

While you are walking, think about your posture. Keep your head up and your back straight. Look around. Notice your environment, including any people. What can you see, hear and smell? Try making eye contact and smiling at anyone you see, even if you really don't feel like it. If you find yourself caught up with negative thoughts like 'I can't do this' or 'I'm so tired' try moving your attention back to everything around you. Walk briskly – faster than you feel like moving ...

After 5 minutes rate your mood again. How is it now? How would it have been after those 5 minutes if you'd spent them sitting alone, brooding on your low mood? The chances are that you'll find your mood has improved a little, or at least not become lower, which is what might have happened had you not got moving.

When our mood is low we naturally stop doing things we usually enjoy because we don't *feel* like doing them. But the less we do, the less we feel like doing. Consequently we feel worse. So we do less still, feel even worse – and down the vicious spiral we go.

This is a basic principle of CBT for low mood. We need to look at our patterns of activity, reinstating things we have stopped doing or are avoiding. Sometimes this can involve a lot of effort – we may no longer enjoy things which we used to get a lot from, and simple things might take far more energy than before.

Let's look at what happens to Linda when her mood becomes low and she stops doing things:

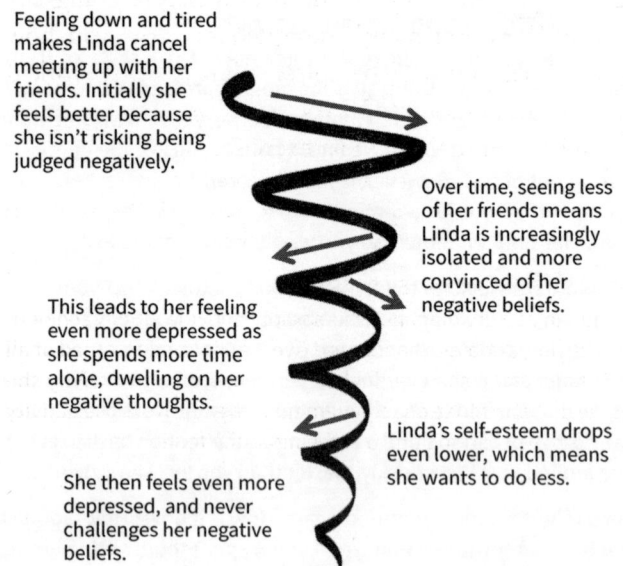

Feeling down and tired makes Linda cancel meeting up with her friends. Initially she feels better because she isn't risking being judged negatively.

Over time, seeing less of her friends means Linda is increasingly isolated and more convinced of her negative beliefs.

This leads to her feeling even more depressed as she spends more time alone, dwelling on her negative thoughts.

Linda's self-esteem drops even lower, which means she wants to do less.

She then feels even more depressed, and never challenges her negative beliefs.

You can see how Linda is caught up in a negative spiral of thoughts and behaviours which maintain and worsen her low mood. Linda's really stuck!

Using behavioural activation

So, how do we go about breaking this downward spiral? The answer is quite simple – start doing more. This is very easy to say and can be very difficult to do. However, if you approach this in a gentle, structured way you can gradually begin to reverse the spiral's direction, travelling up instead of down. For a few days, try monitoring your activity and your mood. Use the mood scale of 0–10 you used earlier. But just before you move onto this next exercise, hold on a moment …

Um, you did go out for that 5-minute walk, didn't you? If you didn't, why not do it now before going any further? The experiment's designed to show you how you really can reverse the spiral – but it won't work if you don't do it. Go on! Off you go!

Right – on with the next exercise.

try it now   ACTIVITY DIARY

Draw up an activity diary like this sample of Linda's. For each period of the day Linda rates her mood (where 0 is not depressed at all and 5 is the worst she ever feels), her pleasure in the activity she is doing and the sense of achievement she gains from the activity (again, where 0 is none and 5 is the most she feels).

Monday		
Morning	Got up, showered, dressed, breakfast	Mood 4/5 Pleasure 0/5 Achievement 1/5
Lunch	Lunch and shopping with friend	Mood 1/5 Pleasure 4/5 Achievement 1/5
Afternoon	Housework, washing	Mood 3/5 Pleasure 1/5 Achievement 4/5
Early evening	Walked dog	Mood 2/5 Pleasure 3/5 Achievement 4/5
Supper	Not hungry – had a snack	Mood 3/5 Pleasure 2/5 Achievement 0/5
Later evening	Completed jigsaw puzzle	Mood 2/5 Pleasure 4/5 Achievement 5/5

Try keeping a diary like this yourself for a week. It might look like a lot of work, but this exercise can provide you with a lot of very useful information about your mood.

After a few days, look at your diary. What do you notice? Most people find that even when they are depressed their mood fluctuates at different times of the day and with different activities. We also know that activities that give us pleasure or use our skills, giving us a sense of achievement, are important in helping us to maintain a better mood.

The next step is to start planning in activities. Use the diary sheet again. Each day plan what you are going to do. Make sure you build in some daily activities which give pleasure and a sense of achievement. These can be very small things – perhaps taking a bath, watching a favourite television programme or completing one small task you'd been meaning to do for ages. Planning and writing down your structure for each day will help you to stick to it, allowing you to monitor your mood and to see how increasing your activity affects it.

The 5-minute rule

When we are feeling really low and lethargic, drained of energy, it can be so hard to get moving. To do anything feels impossible. We can often overwhelm ourselves by thinking about everything that we have to do at once. We then get anxious and often decide there is no point in even starting. Use a '5-minute rule'. Just do something for 5 minutes. Don't focus beyond 5 minutes. Just coax yourself to do this very small thing – then congratulate yourself for doing so. Remember that even just 5 minutes is better than nothing, and when you are depressed it can be the equivalent of an hour's activity when your mood is good. At the end of 5 minutes, see how you feel.

Let's see how behavioural activation helps Linda. Start from the bottom of the spiral and follow the arrows upwards.

Linda starts realizing that there are different ways of looking at her situation and that doing more makes her feel better.

This makes Linda feel more in control and even more positive. It interrupts her negative thoughts.

When the friend agrees, Linda feels just a little more positive.

Linda plans to meet her friend for coffee, and then suggests having lunch and going shopping.

Making this call gives her a sense of achievement. She finds talking to her friend easier than she'd thought.

Linda begins to monitor her activity, and plans to speak to a friend by telephone.

We can see how gradually Linda begins to feel better and become more active. As this happens, her mood improves and the downward spiral is reversed.

USEFUL TIP
Small, gradual changes in your activities can begin to boost your mood. Break things down into small steps – don't expect too much of yourself too quickly.

STEP 2: CHALLENGE NEGATIVE THINKING

Research tells us that when people are depressed their thinking style changes. Experiments comparing depressed thinking with usual thinking show that when depressed we tend to have a negative bias in how we evaluate the world. We see negative events and outcomes as our fault, while ignoring positive outcomes or crediting other people or chance with making them happen. When depressed, it is as if we have put on dark glasses and can only see gloom. Negative thoughts in low mood tend to be automatic; we don't even notice they are there. This isn't really surprising, as the pessimistic nature of the thoughts is perfectly in tune with our low mood. So, everything fits together, meaning we're often not aware of the thoughts, or able to challenge their accuracy. Instead, the first thing we're aware of is that we feel low. To change this, first we need to get better at recognizing these thoughts. Next time you feel low ask yourself: What was going through my mind just before this? You'll find if you start looking, the negative thoughts are there – you just have to recognize them.

COMMON TYPES OF BIASED THINKING IN DEPRESSION
There are a number of ways in which our thinking style may become distorted when we are in a low mood. Let's think about a few of them now.

Mental filter

> *My boss said I hadn't got that report right*
> *– I'm useless.*

When we are low we tend to only notice and accentuate the things which fit with our negative view of ourselves, other people or the world around us. We ignore the fact that our boss also said we had done really well in another project or that we were given an award for our work last year – we concentrate on the things that fit with our depressed view, blind to anything more balanced or positive. Conclusions based on only half the evidence wouldn't stand up in court!

Over-generalizing

> *I can never do anything right.*
> *No-one will ever love me.*

When we are depressed we tend to make global statements about negative events, using words like 'always' and 'never' instead of 'sometimes' or 'right now'. Look out for these words. Then work on collecting *all* the evidence, not just the negative. **Try to recognize that the way things are now is just how they are *now*.** Things seldom stay the same, and we can't predict the future. Nothing is ever *always* the same.

Catastrophizing

> *I haven't finished that project on time. My boss will be furious. I'll lose my job! The mortgage won't get paid. I'll be homeless …*

When feeling low we often allow our thoughts to run away into the worst case scenario before anything has actually happened. We then tell ourselves that such scenarios are inevitable. **Try to take one thing at a time.** The worst doesn't always happen, and catastrophizing makes our mood even lower. Instead, try to think about one thing at a time. You may be able to take measures to prevent the worst happening. Overwhelming yourself with catastrophes, however, means you are more likely to panic and so less likely to take effective action.

Black-and-white thinking

> *I am not perfect at work, therefore I am a failure. My home is not completely spotless, so I'm a slob.*

When we think in 'all or nothing' terms like this we feel inadequate or a failure. The problem with this is that not much in life is really 'all or nothing'. Nobody can be perfect and if we always aim for this we will never be satisfied. **Try aiming for a good job, rather than perfection!** In fact, for minor things like washing the car, experiment with the idea that 'good enough is good enough'.

Mind-reading

> *Rob didn't phone. He thinks I'm an idiot.*
> *Mum didn't say I looked nice. She thinks I'm fat*
> *and ugly.*

Try as we might we can't actually read other's thoughts – a psychologist's job would be much easier if we could! Sometimes we can be reasonably good at reading people, but when we are depressed our negative bias kicks in. We tend to assume we *know* people are thinking negative things. Any ideas why else Rob didn't call? Maybe he was just busy and actually feels bad for not having had time to call, because he likes you. Perhaps Mum did think you looked good but was preoccupied, or just forgot to say so. **Don't jump to conclusions based on biased appraisals.**

Emotional reasoning

> *I feel upset and anxious about work. That must*
> *mean something is really wrong. I am obviously*
> *doing really badly.*

Sometimes the way we feel about something isn't the best guide to how it really is. Try looking at the facts instead of using your feelings as a guide. Ask someone else who isn't experiencing this emotional reaction what they think. **Weigh up the evidence separately from how something *feels*.**

Do you recognize any of these thinking biases? They are very common ways of thinking when we get low in mood, distorting the way we see things, maintaining, prolonging and deepening depression.

CBT sometimes gets accused of being 'just positive thinking'. Actually the way CBT views thoughts is that we should try to re-evaluate them, not making them 'positive', but finding a more realistic, balanced way of viewing and interpreting events. Rarely is any situation either all good or all bad. As Einstein wrote, 'the world ... is a product of our thinking; it cannot be changed without changing our thinking'.

A simple example:

Look out the window. It's pouring with rain.

An overly **negative** view would say:

It's raining. It's never going to stop. I'll be stuck inside forever and I'll never get to do what I want to do outside.

An overly **positive** view would say:

Another lovely day! I'll rush out right now and do what I want to do.

A **realistic** or **balanced** view would say:

> *It's raining, that's disappointing – but it might stop in a couple of hours. If it doesn't I can use an umbrella and still do some of the things I wanted.*

REMEMBER
Watch out for distortions in the way you are thinking. Ask yourself: Is there another way of seeing this? If I wasn't so down right now would I still see it like this?

FINDING BALANCED ALTERNATIVES TO NEGATIVE THOUGHTS

try it now
When you notice yourself feeling low, try to identify the negative automatic thoughts that are driving that mood. Write them down. Now ask yourself the following questions:

- Is there another way of looking at this?

- How might I view this if I was feeling more positive?

- How might I view this in a week's/month's/year's time?

- What would someone else say about this? How might my partner/parent/sibling/friend view this?

- What is the evidence for this thought? Is my thinking distorted or biased in any way? Can I find any evidence that contradicts this thought? What am I missing?

Asking these questions will help you to evaluate your thoughts and to recognize that you may not be seeing things as clearly as you think.

Now, next to the thought you wrote down, list any alternative or balanced thoughts – different ways of looking at the situation which may not be as negative as your original thought.

Let's look at how Linda balances her thoughts:

Negative thought	Alternative balanced thoughts
Losing my job just proves how useless I am at everything.	Losing my job makes me feel useless. In fact there have been other times my work was commended. My last boss considered me very good. This job ended because the company was in trouble. Maybe they really did let me go because I was last to join.

You won't need to write down your thoughts forever, but it is important to practice this as a new skill when you are low. Writing things down in this way helps to develop a new habit of automatically balancing your thoughts, but without practice you'll never get the hang of it.

REMEMBER
Write it down – help yourself learn the new skill of balancing your thoughts through regular practice.

SILENCING THE CRITIC

Lots of us use self-criticism as a tool to spur us on. Over the years we may have learnt that being tough on ourselves can sometimes help to motivate and get us moving when we are flagging. That voice in your head that says 'come on lazy bones – you can't lie around all day – get up and walk the dog/clean the car/call your mother'. We're all familiar with this and can sometimes, in small doses, find it effective. The problem is that when you are low, this critic in your head turns into a bully who no longer has your best interests at heart and isn't motivating, but attacking. Because this bully is part of you, they know exactly what upsets and worries you most and can pounce on those things in a way that just depresses you further.

If you are trying to encourage a struggling friend to do well, which of the following approaches might work best?

> *Come on. You can do this. So you've made mistakes*
> *– that's normal. Focus on what you can do – I*
> *know you can get there if you just keep going.*

OR

> *You idiot. How stupid are you? Look at all*
> *the mistakes you are making! You always get*
> *everything wrong. Go on, try again – give*
> *everyone a laugh.*

It seems obvious doesn't it? Even leaving aside the fact that the second approach is clearly nasty and unkind, which of the two approaches is likely to be more effective in getting the desired result? It seems pretty clear that the first approach is more compassionate and more likely to help someone achieve what they are trying to than the second. Yet when we criticize and beat ourselves up inside our own heads, we're taking the second approach. No wonder we feel more discouraged and depressed.

People who are effective in what they are trying to do usually act as a friend to themselves. They are encouraging, supportive and compassionate to themselves in the same way we might choose to be to a good friend.

Let's look at how Linda might add compassion into her thought balancing:

Negative thought	Compassionate alternative balanced thoughts
I'll never get another job because I'm just useless. I didn't even get shortlisted for that job I applied for last week. There's no point in trying again. I'm just an idiot.	You can't see into the future. Jobs are hard to get in a recession. Yes, you missed out on this one, but it probably had hundreds of applicants. Keep going. You've been successful in the past and so can be again.

A helpful story: the poisoned parrot

Imagine you are given a parrot. This parrot is just a parrot. It doesn't have any knowledge, wisdom or insight – it's bird-brained after all. It recites things 'parrot fashion' without any understanding or comprehension. It's a parrot.

However, this particular parrot is a poisoned parrot. It's been specifically trained to be unhelpful to you, continually commenting on you and your life in a way that constantly criticizes you and puts you down.

For example, your bus gets stuck in a traffic jam, and you arrive at work 5 minutes late. The parrot sits there saying:

*There you go again. Late! You just can't
manage to get there on time can you? So stupid.
If you'd left the house and got the earlier bus you'd have
arrived with loads of time to spare and the boss
would be happy. But you? No way.
Just can't do it. Useless. Waste of space.
Absolutely pathetic!*

How long would you put up with this abuse before throwing a towel over the cage or getting rid of the parrot?

Yet we can often put up with the voice of this internal bully for far too long – decades, even. We hear that 'parrot', believe it, and naturally get upset. This then affects the way we live our lives, the way we behave towards others, what we think about others and the world, and how we think and feel about ourselves.

We can learn to employ an antidote: notice that parrot, and cover the cage!

Say to yourself, 'It's just that parrot again. I don't have to listen to it – it's just a parrot!' Then go and do something else. Put your focus of attention on something other than the parrot. This parrot is poison though, and it won't give up easily. You'll need to keep using that antidote and be persistent! Eventually it will get tired of the towel, tired of you not responding. You'll notice it less and less. It might give up its poison as your antidote overcomes it, or perhaps it will just fly off to wherever poisoned parrots go.

This story was used with the kind permission of Kristina Ivings and Carol Vivyan.

THINKING ABOUT THINKING

Think about your mind. How many thoughts do you think you have in one day? How often do things just pop into your mind that seem random, even strange? Our minds are busy, busy places and not all thoughts deserve our attention. When we are depressed any negative thought that pops into our head tends to grab our attention – we believe it and assume it is true simply because it fits with how we feel. Sometimes this can happen without us even noticing it – negative thoughts follow negative thoughts and our mood gets lower and lower. Training ourselves to become more aware of what is going on in our minds can show us that thoughts are just *thoughts* – not facts. As we have seen, thoughts can be biased, wrong and unhelpful. With practice we can learn to separate the good from the bad and choose which to listen to.

try it now

Imagine you are standing by the side of the road watching the traffic go past. Your thoughts are like big red buses passing. On the side are written the contents of your thoughts. You can choose to get on a bus or just to let it pass. Watch your thoughts. If a bus comes along with 'you are an idiot' written on it – do you *really* want to get on that bus? Will it take you somewhere you want to go? Which bus might be more helpful for you to get on to? Let the 'idiot' bus go past – and get on a better bus.

try it now

Try this exercise, based on a meditation approach called mindfulness.

Sit comfortably with your hands resting in your lap and either close your eyes or gently focus on something in front of you. Now, gradually become aware of the sounds around you – what can you hear? Perhaps a ticking clock, people outside the room or the sound of birds or traffic? Just focus on and become aware of the sounds. Try this for a few minutes ...

What happened in your mind during this exercise? The chances are that it was hard to stick with just focusing on the sounds. Most people find that their minds bombard them with many thoughts, worries and questions. Some thoughts are particularly 'sticky'. They come into our minds and it's very hard to move our attention away from them because of the way they make us feel. Practice this exercise again. Each time you find a thought pulling your attention away, notice that this has happened and simply let that thought go past, like a bus. Move your attention gently back to listening for the next sound. Don't force it – this will only make the thought stickier. Just coax your mind back to what you are focusing on. You may have to do this dozens of times in just a few minutes. That's normal. But the more times you wander into thinking, the more chances to practice tuning back to the sounds you will have. If you find it difficult, watch out for that parrot telling you that you can't do it!

FIND OUT WHAT WORKS FOR YOU

Don't forget, this book offers ideas and guidance using a range of techniques. You are looking to find those that help you most. Some people who suffer from depression only make limited progress with certain psychological therapies, and when such people are also averse to drugs, they can be in trouble. However, there are always ways through.

case study JOE

Joe's childhood was isolated after his mother lost her first child to diphtheria. Deprived of social contact and play, Joe was bullied at school and was a 'dreamer' in class. In his working life he was introverted, tense, serious and hyper-critical of others. He was never happy with his place on the career ladder, although he was considered by others to be very successful. He suffered depression throughout his adult life and only made friends through team sports and sports clubs. He was married for 20 years before separating from his wife.

Joe consulted 6 or 7 psychotherapists over 2 decades. In therapy he learned to understand the possible causes of his depression, but his moods persisted. Indeed, at times they got worse and when he started having suicidal thoughts he consulted a psychiatrist. The various drugs prescribed all had side-effects Joe couldn't tolerate and he felt he was back to square one.

Then Joe came across the Buddhist teachings on suffering and the causes of suffering. He learned to use **mindfulness meditation** to investigate the reality of his own distress. He started to think about the way in which his mind worked and began to recognize the negative thoughts that drove his depression. He found that as he did this he became more able to question the validity of those thoughts and to change how he interpreted things around him. Joe also looked at his activities and lifestyle. He built in more time with people he cared about and tried out some new social activities.

Friends of Joe have witnessed great changes in his personality. He is confident that the dark moods are unlikely to overwhelm him in the future because he now uses the investigative tool of mindfulness meditation to determine the reality of any given problem. The illusions and distortions created by his early childhood do still crop up but he can now see them for what they are and not let them damage his health and relationships.

CBT recognizes the value of mindfulness meditation and it can become a useful tool to aid therapy.

SUMMARY – THE DOS AND DON'TS OF BEING IN A DEPRESSED MOOD

Do!	Don't!
Tell someone how you feel	Beat yourself up
Get moving – do something active for 5 minutes	Stay in bed
Treat yourself as you would a friend	Expect too much of yourself
Examine your thoughts for biases	Listen to that parrot!
Remember thoughts are not facts	

CHAPTER 7

Coping with bad times

*People are like stained glass windows. They
sparkle and shine when the sun is out. But when
the darkness sets in their true beauty is revealed
only if there is a light within.*

Elizabeth Kubler Ross

Chapters 4 and 6 describe how CBT highlights the way distortions in our thoughts and beliefs about events in our life lead to emotional distress and/or patterns of unhelpful behaviour. However, what does CBT say about situations where our so called negative thoughts could be correct? Perhaps you've lost someone you love. It might be accurate to say, 'I'll never see that person again.' That's no exaggeration, or other type of negative distortion – anyone, understandably, would be extremely upset. We all experience unpleasantness in our lives sometimes, but some people are faced with more, and tougher, things than others.

Think of any upsetting things you've experienced in the last five years. Write them into the following table. If need be, draw a bigger table.

Date	Event	Feelings

Here are some events that within Western society have been shown to be potentially the most stressful. Have you experienced any of them?

- Death of a spouse, partner or close family member

- Prison sentence

- Death of a close friend

- Divorce/family break-up

- Relationship break-up

- Unwanted pregnancy/miscarriage (you or your partner)

- Period of homelessness/housing problems

- Significant debt/financial problems

- Unemployment

- Serious illness of you or family member.

Are there any others on your list? Do fill in the table and take the time and effort to write down your feelings too.

If you have experienced one or several from our list, research suggests you are more likely to experience both physical and mental health problems, and to find yourself less able to cope socially. The more you've experienced, the worse you are likely to feel.

Actually, it's not quite as simple as this. Many variables dictate a person's responses to stress. Some people are more resilient and therefore cope better with stressful events than others. While certain factors like biological vulnerabilities, upbringing or historical events are beyond our control, there are many things we can do to improve our chances of coping more successfully with stressful events. Contrary to popular belief, some people aren't just stronger than others. Our resilience and ability to manage stress changes throughout our lifetime, and normally 'strong' people can at certain times find themselves feeling defeated by things which they coped well with at other times.

It's helpful to think we all have a limited capacity for stress, rather like a bucket which holds a finite amount of water. We all experience some stress in everyday life. Indeed *some* stress is needed to motivate and energize us into action. However, if our stress bucket is consistently almost full, it won't take very much more for it to overflow, or for us to develop difficulties in coping. This is why sometimes a seemingly small problem coming on top of many other comparatively small problems can finally make us feel we're falling apart. We've perhaps coped very successfully with everything until then, but that's where it all ends in tears.

Research shows that many factors dictate how we are affected by negative life events, both big and small. Some are discussed briefly in a moment. Read each one and think about your own life experiences. How have they equipped you to cope with stressful life events? Are there particular ways in which you are vulnerable to stress, or in which you are more resilient? Negative experiences can often work in different ways for different people at different times. Sometimes we can learn from things and become more skilled through our experiences. At other times we can't and the identical negative experiences can have a much greater effect. If you find yourself less able to cope successfully with stress, there are probably many good reasons for this. It's certainly *not* that you are just weak.

KEY FACTORS DETERMINING COPING

- Meaning of the event
- Identification of your strengths and abilities
- Historical factors
- Personality traits.

Let's look at these in a bit more detail.

The *meaning* we give to events can change both our reactions, and our ability to cope with them. For example, if we think we caused a problem by doing something wrong, our reaction will depend on whether or not we view our mistake as understandable. If we think the mistake was forgivable, we may be able to learn from it and make positive changes for the future. But if we judge ourselves as 'bad' or flawed we may feel helpless to change and learn from the experience. One view means we can still feel good about ourselves, the other means we don't.

If we think we have the **strengths and abilities** to cope with a situation, and that it is manageable, however awful it may be, then we'll use strategies which are more likely to have a positive outcome. These are **approach-related strategies** like problem-solving, learning from difficult experiences, and using support from others.

But if we think that we can't cope, or that the problem is totally unmanageable, we're more likely to use less successful strategies. These **avoidance-related strategies** include staying away from others, cutting down on the things we normally do, using drugs or alcohol to escape or pretending that the problem doesn't exist, while vainly hoping it will simply disappear. Predictably, these strategies have been shown to be less successful.

Historical factors like upbringing, education and early life experiences are also important in coping with stress. If we've been taught that showing emotion means we're weak, we may be angry with ourselves for what are normal reactions under the circumstances. So for example, if we lose a very dear partner or friend, we may cope less well as we berate ourselves for feeling very normal emotions such as sadness and grief. But if we've been taught that we should express and share our emotions, though our grief will still be as strong, it will be easier to manage and we'll cope better with it.

Personality traits such as optimism or sociability may also make us more or less vulnerable to stress. Social support has been shown to be very important in how we manage after a stressful life event. People who are naturally sociable, or form close relationships therefore have an advantage here.

Many other factors have a role in how well you cope. These include your age, socio-economic status or life-stage. Clearly some factors above are more within our control than others. The meaning you give to events and your prediction of how you'll cope are central to CBT. By the time you are finished with this book, you are likely to have learned to both identify and to handle these in very different ways.

As for the influence of historical factors and personality traits, the good news is that none of these are insurmountable. No matter what the situation, we can all learn to manage stress better, irrespective of our background, gender, age or experience.

GOLDEN RULES FOR COPING WITH STRESSFUL LIFE EVENTS

In the following sections we will look at the golden rules for coping with a stressful situation and looking after yourself during a difficult time.

1. TAKE CARE OF THE BASICS

When bad things happen it's tempting to curl up into a ball under the duvet and get out of your usual routines and habits. Taking care of your basic needs is even more important at these times. You may not feel like eating, and sleeping may seem impossible, but it's vital you take care of yourself. Eat little and often rather than trying to force down normal meals, but make sure you do eat – emotional times can drain us of energy and our bodies need food, despite our minds insisting we're not hungry and don't want it. Try to rest even if your sleep is disturbed. Many people find that sleeping tablets used sensibly and for as short a time period as possible can be a useful way to get through the first few nights after something difficult has happened. If you are having problems sleeping, take a look at Chapter 4 for some handy tips.

2. KEEP (REASONABLY) BUSY

Going through the motions of our usual routines can actually be very helpful during difficult times. If you usually take the dog for a walk in the morning or pop out to the local shop for a paper in the afternoon, try continuing with these activities. We all have familiar tasks which can help keep us in touch with normal life and we do almost on autopilot. They can prove soothing at difficult times, and can also remind us that our life is still going on and that we can still control at least some aspects of it. Sometimes we have to go through the motions in order to then move on with life.

case study BHAVEN

Bhaven, a recently widowed young husband, described how about a year after his wife died he went to a party. He went because his friends had asked him and he felt bad about declining yet another invitation. He didn't particularly want to go and when there said that he did not really enjoy himself. However, Bhaven admitted afterwards that it had not been quite as bad as he'd anticipated, and was even able to think that next time it would be easier to go, and that eventually he might even be able to start enjoying parties, like he used to before his bereavement.

Beware – sometimes people use work or other activities as a way of hiding from or avoiding their emotions. This can mean it takes longer to recover.

3. EXERCISE

This could well be the last thing you feel like doing. However, there's good evidence that physical exercise is important for managing low mood and stress. Even just a brisk walk round the block or in a local park may improve your mood just a little and will probably have a more positive effect than simply sitting and focusing on your problems. SO … the exercise is … EXERCISE! Get moving! Go on, force yourself to do anything physically active – just for 5 minutes. Do it!

4. ALLOW YOURSELF TO FEEL SAD

This may sound strange, but many people spend a lot of time fighting normal, natural reactions to what has happened to them. Losses and disappointments can bring grief. Feeling and expressing grief and sadness is neither weak nor pointless. It may even be the only way we can truly heal and move on. Everyone experiences and expresses emotions differently. No way is either right or wrong. However, when we struggle and fight against feeling painful emotions, usually we only make them worse.

Think of what you might do if you were trapped in quicksand. Your first instinct would be to struggle desperately to escape. But that's just the opposite of what you should do. The more you flail around, the faster the sand will suck you down. Your best bet is to stop struggling, lie flat and slowly inch forward. It's the same with painful emotions. Stop struggling and try to just tolerate the pain. Stay with it. Don't fight the feelings – they'll get stronger and try to drag you down. Remind yourself that what you are feeling is normal, natural and understandable. Most importantly, remind yourself that, with time, you will heal and the feelings will pass. That does not mean that your loss will ever be either forgotten or even totally eradicated, but just that your feelings will become less intense and less painful over time. Some things will always hurt, but they will gradually have less effect on your ability to function – even if, while you are going through it, that time seems impossible to imagine, or a really long way off.

When you feel sad, just allow yourself to feel that way. Remember:

- You are feeling this way for a reason

- This is just how you feel right now

- There will be good days and bad days

- Feelings change and even grief comes and goes in intensity

- Make the most of the good days and on the bad days be good to yourself

- Do things that soothe and comfort you and be gentle with yourself

- Treat yourself the way you'd treat a close friend or a child who was in pain

- You will heal faster if show yourself kindness

- Don't scold yourself or tell yourself 'you should be over it by now'

- Be your own best friend.

Finally, and most important of all:

- Do what is right for you.

It's utterly useless demanding that you 'pull yourself together'. If it were that simple, you'd have done it ages ago – and we wouldn't be writing this book!

USEFUL TIP: HOW TO BE GENTLE TO YOURSELF

Take a warm bath. Have a massage. Eat comfort food (bananas and chocolate can help – but don't overdo it!). Make time to talk to a friend. Watch a favourite film. Go for a walk in the country-side. Buy a new outfit.

See how many other ideas you can come up with which we haven't mentioned.

We aren't suggesting for a moment that any of these things will take your emotional pain away completely. However, they may help to relax you a little. Indulge yourself just enough to give yourself space to overcome the numbness and start to feel again.

---

5. WATCH OUT FOR DISTORTED THINKING

At the beginning of this chapter we warned how *some* negative thoughts at difficult times in our lives are inevitable, and may actually be accurate. But that doesn't mean they *all* are. We can still have distortions and mis-interpretations in our thinking at times like this. Notice the things that are going through your mind. How accurate are the assumptions you are making and the things you are saying to yourself right now? How helpful are they to you?

case study JENNY

Jenny is going through a divorce, having found out her husband has had a series of affairs. He has finally left her for a young-er woman he met at work. Naturally she is very angry, sad and distressed. Many thoughts go through her mind. Some of these thoughts, such as 'he doesn't love me anymore', 'he finds her more attractive than me' or 'he wants to be with her more than with me' may well be true. Of course they'll hurt terribly. However, these thoughts are accompanied by many others which are less true and just deepen distress unnecessarily. Examples are given in the table below.

Distorted thought	Logical counter-arguments
No-one will ever love me or find me attractive again	
Every man will eventually leave for a younger woman – they're all the same	
I will never be happy again	

See if you can find some logical counter-arguments to these examples of distortions in Jenny's thinking. Write them in the right hand column above. We've given some suggestions in the table below, and of course there can be a whole lot more.

Distorted thought	Logical counter-arguments
No-one will ever love me or find me attractive again	None of us can predict the future
Every man will eventually leave for a younger woman – they're all the same	Everyone is different
I will never be happy again	Painful feelings rarely last forever, even if they are excruciating at the time

Though the thoughts in the left hand column may *feel* as true as Jenny's initial thoughts ('he doesn't love me anymore', 'he finds her more attractive than me,' and 'he wants to be with her more than with me') an outsider can see how they are distortions of reality.

Look at the examples of thought distortions that are listed in Chapter 6. Do any of these apply to you in your current situation? Would it be helpful to use some of the thought balancing techniques suggested in Chapter 6? Or try drawing up your own table like Jenny's above. Either way, see if you can identify any distorted thoughts you've been having, and work on balancing them or providing logical counter-arguments.

REMEMBER

However awful or painful your situation is, it may be that not all your very negative thoughts or predictions are completely true. Work on catching yourself if you are trying to predict the future or jumping to conclusions based on your current situation. Don't give yourself more pain by telling yourself that you can never move on or get over this.

Have you ever noticed how in a city it can be very hard, if not impossible, to see the stars in the night sky? This is because the light pollution generated by buildings, streetlights and cars prevents us from seeing the light of those stars. The wonderful thing is that the stars are still there. We just can't see them. As soon as the lights are turned off or we go out into the countryside we see the stars again. It is a little like this with good things in our lives when we are in pain. We can't see any positives. However, it is so important to remind ourselves that they are still there somewhere. When our situation changes and the painful feelings dim even just a little, then the positives start emerging and become clearer again. None of this is about denying or avoiding the negatives. They are all too real. It is about recognizing that nothing is ever either pure black or white.

6. CUT DOWN ON SELF-CRITICISM

Telling yourself that you are weak because you are 'not coping' is not likely to be helpful right now. Check out the story of the poisoned parrot in Chapter 6. Recognize this? Be a friend to yourself. You can be firm and encourage yourself to move forward – but at the right pace and in a supportive, kindly way. You will find that this is much more effective than beating yourself up for reacting in a very normal way, which you would probably understand and forgive very easily in others.

7. LEAN ON OTHERS

We all need support in difficult times. It can be very hard to admit that we need help, or are not coping as well as we'd like. It can make us feel that we are weak or useless, and that people whose opinion we value think less of us. However, very often when we make the first move to ask for support, we can be pleasantly surprised. Usually people are pleased to be asked to help. We all like to be needed or feel useful and sometimes helping someone else can make us feel better about things in our own lives. Give those around you a chance. Reach out and ask for help – even in small, practical ways. You'll probably be surprised with the results. Be wise in your choices – select people you think will be supportive, and let them know what you need. If you can be brave enough to communicate your feelings and needs, the chances are those needs will be met. If help isn't forthcoming, try to let go of any anger about this – you have enough to deal with right now.

8. WRITE IT DOWN

There is evidence that writing about negative events can help people to feel more positive, and can even reduce the number of physical ailments they experience in the months after a difficult life event.

try it now

Try this exercise. Set yourself a short time every day over the next few days to write about your experiences, feelings and reactions. You can write about the same thing every day or something different each day. Don't think too closely about what you write or worry about punctuation, spelling, grammar or even accuracy! Just write. Then get rid of that writing. Don't re-read it. Throw it away, burn it, recycle it, rip it up – whatever works for you. This exercise is designed to help make sense of and process emotionally charged information, which then helps us to move on and be less affected by stress related to the experiences we have written about. It also might explain why diary keeping has always been so popular.

WHEN DO NORMAL REACTIONS TO STRESS BECOME A MENTAL HEALTH PROBLEM?

It's a fact that sometimes adverse life events can trigger mental health problems such as anxiety or depression. It's very difficult for professionals to say exactly where 'normal' reactions stop and a mental health problem begins. For some people adjustment to an adverse life event or loss takes a very long time – it may be months or even years before you feel better or able to fully move on with life. In CBT it's often recommended that people don't get active or formal treatment in the first few months after a negative life event because distress – even depression and anxiety – is a very normal reaction to an abnormal event in a person's life. However, it can also be true that some people, for whatever reason, get very stuck following a negative life event and find it impossible to move on or to rebuild their lives.

If you feel that this has happened to you, think first about whether you are perhaps expecting too much, too soon of yourself. Seriously, is it realistic to expect to be 'over it' by now? If you do feel you may be depressed or anxious then Chapter 3 and Chapter 6 could help you find some ways to tackle this. Talk to people around you. Do they think you are depressed? Sometimes others can judge this much better than we can ourselves. Most importantly, if you feel that things are out of control and the self-help strategies suggested here aren't working, despite time and practice, then ask for help. What does your GP think? Chapter 10 outlines the many sources of help available.

POST-TRAUMATIC STRESS DISORDER (PTSD)

So far in this chapter we have talked about quite common, but difficult and stressful life events. These can feel traumatic, but the traumatic events we are looking at in this section are in a different league. The kind of events which might lead to developing PTSD are ones people believe are life threatening either to themselves or to someone close to them, and that they are helpless to do anything about. Traumatic events, while rare, can happen to anyone. Reading newspaper stories can make one wonder how people cope with situations involving serious injury or death. PTSD arises when the normal responses to an abnormal event:

- Begin within 6 months of the event or period of trauma

- Carry on for more than 3 months after this time

- Start over 6 months after the trauma – this is called delayed onset PTSD.

People experiencing traumatic events feel intense fear, helplessness or horror. Traumatic events are *outside our normal experience*. Divorce, bereavements (except those caused by traumatic events), loss of employment, chronic illness and marital or domestic conflict therefore don't count as trauma even though, as we have seen, they may be triggers for extreme stress and even anxiety and depression. PTSD has a very particular set of symptoms, different to those of mental health problems, though PTSD sufferers may also have elements of both depression and anxiety. Examples of common traumas that may lead to PTSD are road traffic accidents, attacks or assaults, combat situations and being caught up in natural disasters or acts of terrorism.

Commonly, after a traumatic event most people get some of the following reactions:

- Distressing thoughts, memories, images, dreams or flashbacks of the trauma which keep recurring – sometimes there are blank bits which the person can't recall

- Avoidance; where possible, you try to avoid places, people, activities, thoughts, feelings, conversations or anything else which might trigger memories or thoughts of the trauma

- Emotional numbness, detachment from others, difficulty having loving feelings

- Seeing the future as hopeless, pointless and likely to be short-lived

- Losing interest in and stopping activities you used to enjoy

- Increased arousal – you are easily startled and don't like bright lights or loud noise

- Problems sleeping, irritability, anger, difficulty concentrating and increased vigilance.

These reactions may be very normal at first and usually pass with time and the use of some good coping strategies. However, for some people, for reasons that we don't yet fully understand, these symptoms don't improve and may even worsen over time. Have you gone through a traumatic event and been experiencing any of the feelings above? If they aren't improving, or are even worsening, you may find the exercises later in this chapter useful. But if things get no better, or even get worse, then it's very important to seek professional help, either from your doctor, or a qualified mental health professional. Guidance on where to find such help is given in Chapter 10.

WHY DO FLASHBACKS HAPPEN?

We don't know exactly why the brain re-experiences things in the way that it does following trauma. Psychologists believe, however, that it's to do with the way in which the brain processes events and stores them as memories. Just imagine your memory is like the linen cupboard of a neat, organized housewife (or husband). Each item is first sorted, neatly folded, then put away in an ordered fashion. Sheets go with sheets in one pile, pillow cases are all together in another. When the door is shut, our memories stay locked up inside. When something happens and we choose to open the door, we can take out a memory and examine it or use it. Occasionally things fall off the shelf and out the door at strange or inappropriate times, but generally we can pack stuff away again and it's no big deal.

Now imagine that along comes a huge, irregularly-shaped duvet which has to be put into the cupboard. It doesn't fit. It won't fold into a neat, organized shape and no matter how our heroine (or hero!) tries to pack it away, it just keeps falling back out, forcing the door open. This is what seems to happen with traumatic memories. Our brains initially seem to find it impossible to make sense of or properly process these types of memory – perhaps because they are so far out of our normal experiences and expectations of life. There's no template to fit them in. It's as if our brains need to keep re-experiencing the memory in order to try to process it – to pack it away. All the time it's not processed in the usual way, we experience the event not as a normal memory from the past but as a new event – *just as if it is happening right now*. There is some evidence that the part of the brain which is associated with traumatic memories is the same part connecting to our *flight, fight or freeze* mechanism which we discussed in Chapter 3 when looking at anxiety. So all our anxiety reactions are triggered each time this event pops up, unbidden, into our minds. Flashbacks can be terrifying, horrible experiences, but they may actually be our brain's way of trying to heal itself. Fortunately there are ways of lending your brain a helping hand to accomplish this.

WHAT YOU SHOULD *NOT* DO

Quite often people use alcohol or even illicit drugs to try to relax, to help them sleep and to stop thinking about the event. Unfortunately this usually proves at best unsuccessful, and at worst, can increase your problems by adding alcohol or drug dependency to them. If you try to drown your problems with alcohol, chances are they will learn to swim!

WHAT YOU SHOULD DO

Let yourself accept the whole array of feelings that you are experiencing. Remind yourself that these are perfectly normal under the circumstances. They don't mean you are 'going crazy', 'being pathetic', or any similarly unhelpful name-calling you are doing.

Now look at what you are doing differently compared to before the trauma. Are you over-cautious? If you were attacked, are you now unwilling to leave home after dark, despite living in a safe area? If what you are doing really is excessively cautious compared to before, make a list of the things you are now avoiding and start facing them one at a time, perhaps trying the easiest first. You might even ask a friend or relative to accompany you in the first instance, but then move on to doing it on your own, just as you did before. At first it can be very frightening but as you repeat the actions, and find that nothing awful happens, you'll probably find your self-confidence steadily increasing.

It's so important that you do re-evaluate the actual event, perhaps talking it through with someone. Maybe there really is something to be learned. In the example of a violent burglary, you might decide to install a burglar alarm and use it for certain rooms even when you are in the house. If you've had a car accident linked to driving in bad weather conditions, you might decide to do an advanced driving course. Work on discriminating between what is a reasoned, more cautious approach, from the probably quite extreme course of action being dictated by your anxiety alone.

SEPARATING THEN FROM NOW

The way our minds store information after a trauma is often a bit jumbled up. At the time of the trauma, everything tends to happen really quickly, and each bit can get tangled up with the others. Quite often an innocuous smell, sound, or even piece of music can take you back into the trauma, which then triggers a complete flashback of the whole event.

If you do experience this, then remind yourself kindly but firmly that that was *then*, but *then* is not *now*. Tell yourself where you are, for instance in a different car, and remind yourself of the date today, and the date of the trauma, to help you separate the then from the now. You can apply this technique to any trauma and it is especially helpful if you are experiencing frequent flashbacks.

COPING WITH FLASHBACKS

Tell yourself you are having a flashback and that this is OK and very normal in people who have experienced trauma.

Remind yourself that the worst is over – it happened in the past but it is not happening now. Remember: 'that was then, and this is now'. The traumatized part of you is giving you these memories to use in your healing and, however terrible you feel, you survived the awfulness then, which means you can survive and get through what you are remembering now.

Call on the stronger part of you to tell the traumatized part that you are not alone, not in any danger now, and that you can get through this. Let your traumatized self know that it's OK to remember and to feel what you feel, and that this will help you in your healing and getting over what happened to you. However hard it is for you, your brain is attempting to heal itself the only way that it can.

Try some of these ways of grounding yourself by becoming more aware of the present:

- Stand up, stamp your feet, jump up and down, dance about, clap your hands, remind yourself where you are *now*

- Look around the room, notice the colours, the people, the shapes of things – make it more real

- Listen to and really notice the sounds around you like traffic, voices, machinery or music

- Notice the sensations in your body, the boundary of your skin, your clothes, the chair or floor supporting you

- Pinch yourself or ping an elastic band on your wrist – that feeling is in the *now*. The things you are reexperiencing were in the past.

Take care of your breathing. Breathe deeply down to your diaphragm; put your hand there (just above your navel) and breathe so that your hand gets pushed up and down. Imagine you have a balloon in your tummy, inflating as you breathe in and deflating as you breathe out. When we get scared, we breathe too quickly and shallowly and our body begins to panic because we're not getting enough oxygen. This causes dizziness, shakiness and more panic. Breathing more slowly and deeply will stop the panic.

If you have lost a sense of where you end and the rest of the world begins, rub your arms and legs so you can feel the edges of your body, the boundary of you. Wrap yourself in a blanket and feel it around you.

Get support if you would like it. Let people close to you know about the flashbacks so they can help if you want them to. That might mean holding you, talking to you and helping you to reconnect with the present, to remember you are safe and cared for now.

Flashbacks are powerful experiences which drain your energy. Take time to look after yourself when you have had a flashback. Try a warm, relaxing bath, have a nap (not both at once!), have something hot to drink, play some soothing music or just take some quiet time for yourself. You deserve to be taken care of given all you've been through.

When you feel ready, write down all you can remember about the flashback and focus on how you got through it. This will help you to remember information for your healing and to remind you that you did get through it (and so can again).

Remember you are not crazy – flashbacks are normal and part of healing.

This section has been used with the kind permission of Carol Vivyan.

PLAN SOME ACTIVITIES

After a trauma, especially one involving loss, everything can feel both overwhelming and yet at the same time pointless. It's tempting to do very little. But the catch is, the less you do, the worse you feel and the more everything mounts up. So you feel still worse, do even less, watch the backlog grow and just don't know where to start – and anyway, you think, what's the point?

A way out of this hole is to start with a rough plan for each day – one thing to do in the morning, one in the afternoon and something for the evening. Try to plan one week ahead, carry out what your plan tells you to do and then plan the next whole week. Congratulate yourself for having done the things you achieve, but work on not beating yourself up for what you didn't do. Instead, break that activity down into smaller steps and put them onto next week's schedule.

MOVE FROM AUDIENCE TO FILM DIRECTOR

Flashbacks can be horribly unpleasant and even terrifying. One way to deal with them, as we've mentioned above, is the technique of separating *then* from *now*. Another solution is to discover that you have a level of control that maybe you thought was impossible. While you are unlikely to prevent flashbacks, when they do start, you can experiment with treating the scene as a film you are watching. Instead of being a passive member of the audience, take over the role of film director. Start modifying the flashback, treating it as if it were a film, rather than the real event. Tell yourself that though the event did happen, what is now going through your head is something your mind is picturing. Give yourself the option that you really don't have to watch that particular picture again, and let yourself instead work on developing a different ending. Some people change the distance from the event, so it becomes small and far away, others might take the aggressor and make them look ridiculous, for instance by imagining them in a clown-suit, wearing clown's make up, and walking along on their hands. Still others bring someone or something comforting into the event, so they feel they are receiving warmth and support. The important thing with this technique, called **imagery re-scripting**, is that *in no way are you denying that the actual event took place*. What you are doing is being creative with your imagination. There is no reason why you have to keep reliving it *as it was*. You discover that, surprisingly, you do have a choice of what you watch inside your mind, and you *can* choose something that feels better in some way.

THE SILVER LINING

Finally, it is worth remembering that many people who have recovered from a traumatic or stressful life event have said that with hindsight, they now see it as an important wake up call. They then start making many changes to their world which they otherwise would not have done. In fact, as one author put it 'the individual is able to see him or herself as stronger, wiser and with a new value to his or her life'. Some researchers call this **post-traumatic growth** and think that to achieve the optimum level of growth and learning in life a certain amount of adversity may be important or even essential. It can be very hard to believe or recognize this when in the midst of a crisis, and even reading this may make you think we're being insensitive or patronizing. However, it's worth remembering that this idea has been around in the writings of philosophers for a very long time. Nietzsche is quoted as saying, 'What does not destroy me, makes me strong'.

REMEMBER
Above all, remember – and heed – the words of Winston Churchill:

If you're going through Hell, keep going.

CHAPTER 8

Maintaining progress and reducing recurrences

*If you can find a path with no obstacles, it
probably doesn't lead anywhere.*

Frank A. Clark

You've reached the penultimate chapter. Hopefully there is quite a difference between how you view and feel about things now compared with when you started this book. However, it's an unfortunate fact that many of the perceptions and feelings we have discussed can return – at least for a short while. Difficult times will always happen. Research has found, for example, that if you've had more than one episode of depression then you are quite likely to experience depression again. So, the million dollar question is …

What can you do about it?

case study AGATA

Agata is in her early 30s, with two children under the age of five. She experiences depression after the birth of her youngest child, and successfully works through this using many of the strategies we've discussed. Agata recognizes her thinking is very negative and that she's constantly berating herself for not being a sufficiently good mother. She realizes that much of this stems from her longstanding fear of failure. Agata identifies her negative thoughts, weighing up the evidence for them, instead of just assuming they're true. So, instead of assuming she *knows* how things are in other families, she tests out some negative thoughts by observing what other new mothers do and asking some of them about their own experiences. Agata also practices challenging the self-critical voice in her head, treating herself the way she'd treat a friend who was feeling down. She shares her feelings with her partner, asking him to help her more, which gives her some time for herself. Noticing that she's stopped doing things she previously enjoyed, like seeing her friends and exercising, Agata starts making time to see her closest friends and also joins an exercise class for mothers, which she can fit in around the children. Agata begins to feel better, and life returns to how it was before the birth of her second child.

However, Agata is still bothered by thoughts that she's not good enough, and isn't measuring up to others. Arguing with her thoughts is hard work, so she resorts to saying to herself 'just stop being silly' and 'pull yourself together'. Then Agata's husband is unexpectedly made redundant and things become stressful at home. Agata feels powerless to help him, and goes back to seriously berating herself for not being good enough. Feeling unable to ask her husband to continue helping her whilst he's stressed, she once again cuts down on things she's doing for herself, like seeing friends and going to her exercise classes. Her husband finds a new job, but his salary's lower, and the hours longer. Agata continues limiting her activity to just caring for him and the children. She notices she's starting to feel tired and unhappy and – you guessed it – now begins beating herself up about that – 'Here I go again … I'm so useless, being depressed and pathetic when I should be grateful I have healthy children and we have money coming in again – I'm such a loser!'

Research shows that relapse or recurrence of depression, and indeed other psychological problems, can happen because even though we're feeling better, frequently we still keep some of our old, unhelpful ways of thinking and behaving. This means we're quite vulnerable when life becomes hard or stressful. The old patterns kick in and down we spiral … again. While many of us look after our physical health, knowing all about diet, exercise, check-ups and screenings, we're not so aware of our mental health. It's equally important to work on our mental health when we're well as when we're not. Prioritizing our mental wellbeing and learning to keep ourselves mentally 'fit', means we're less likely to spiral down into depression or develop anxiety-related problems when life gets stressful. Think of it as a little like losing weight. If you really want to lose weight and stay trim, crash dieting and then going back to eating exactly the same way as before is useless. You'll very quickly regain your former size. If you really want to lose weight and stay in shape you have to permanently change the way in which you eat. The same is true of maintaining your improved mood and mental health. To help you do this, we've summarized a number of ideas which can increase your resilience and reduce your vulnerability to ongoing mental health problems.

REMEMBER
It is as important to pay attention to maintaining your mental health as your physical health!

KEEP UP GOOD HABITS

We hope this book helps you develop many useful strategies to manage your mood, mental health and indeed your whole life. All the ideas you find helpful can be applied to everyday life, even when you are not feeling particularly down or anxious. The strategies discussed in

Chapter 3 and Chapter 6 can be used to develop habits to keep you feeling well rather than just being used when you are struggling.

Take activity scheduling, for example. It's important that we all ensure that our weekly schedule includes pleasurable activities or those which provide a sense of achievement. If you build this into your normal week, you'll be less vulnerable to problems when things are stressful or challenging. Equally, watching out for distortions or unhelpful thought patterns and being aware of the ways we interpret the world, even when we are well, can increase our future resilience.

USEFUL TIP
As you begin to feel better it can be really easy to drop new good habits because you don't need them as much – try to maintain them in order to keep your mental health in good condition.

WHEN EVERYDAY BLUES BECOME TOXIC – WHAT IS NORMAL?

Everyone has down days. It's perfectly normal to feel sad, lonely or blue at times in our lives. Sometimes we know exactly why we feel that way, sometimes we don't. Unfortunately, if you have experienced depression, you can sometimes feel very frightened by even 'normal' low mood. You find yourself feeling a bit low or flat, and start panicking that this means you are becoming depressed again. This in itself can trigger a cycle of negative thinking which actually worsens your mood. Thoughts like 'here I go again, I'm never going to get over this' and 'nothing

I do makes any difference – I'll always be depressed' lead to unhelpful behaviours like avoidance or withdrawal, as we've described in Chapter 6. This then triggers further negative thoughts and unhelpful behaviours, and before you know it, the downward spiral has begun again. It is really important to allow yourself to have low times without panicking or immediately labeling them as depression.

Remember these key points:

- It wouldn't be normal to never be sad

- This is just a difficult time

- Get active – even if you don't feel like it – do anything that might lift your mood

- Don't give in to the urge to shut yourself away and brood – do the opposite

- Talk to someone

- Curl up on the sofa and escape into a good book or a film

- Do anything which takes you away from ruminating on negative thoughts

- Remind yourself that going over your worries or negative thoughts right now won't be helpful – you don't come up with your best solutions while you are feeling down

- If you need to take some time out and have a 'duvet day', that's okay now and then, but don't let it become a habit

- Don't beat yourself up for anything you think is lazy or frivolous

- Recognize and try to meet your needs

- Be kind to yourself.

Sometimes when someone has been depressed the people around them can also assume they are becoming depressed again if they have a low day. Talk to your loved ones about this – they need to allow you the normal ups and downs of life without panicking that you are once again depressed.

SPOT YOUR EARLY WARNING SIGNS

When we have recovered from a period of depression it's often very tempting to want to move on as quickly as possible and not think about what happened. However, learning from what's happened can decrease the likelihood of recurrence. Think back to when you first became depressed. The chances are that it built up gradually over a period of time. Ask yourself the following:

- What happened around that time?

- What sorts of things were going through your mind?

- What did you stop doing?

- What did you do more of?

- What did other people around you notice?

Really think this through. Talk to those close to you – they may well have noticed things you overlooked. Now write yourself a plan. What is it that you have to watch out for? What should you do if you notice early signs that all is not well with your mood? Doing this can enable you to head off future problems. Tackle problems early, before they get out of your control. Notice your urges to do things you now know are unhelpful to you, like avoiding social events or dwelling on sad memories from the past. Help yourself not to act on those urges – remember that the thing you feel like doing (like staying in bed or cancelling all engagements) might not be the right or most helpful thing for you to do. Let someone else in on your plan – ask for help in detecting problems and in heading them off before they seriously interfere with your quality of life.

EXPECT SETBACKS AND LEARN FROM THEM

Recovering from any psychological problem is never a smooth process. There will always be setbacks and days where you feel as if you are back to square 1. Remind yourself that it isn't really so. However bad you feel right at that moment, it cannot undo what you have learned from the progress you've already made. If you did it once, you can do it again, even though it may seem really difficult at the time. Try seeing setbacks as an opportunity to learn more about staying well. Don't jump to conclusions that this means nothing is working and all is lost – imagining catastrophes before they have happened is just counterproductive. Instead, try identifying what tripped you up this time. How could you avoid this kind of thing in the future? What will you need to do differently?

PRACTICE BEING 'MINDFUL'

In Chapter 4 we introduced the concept of **mindfulness**. Mindfulness has become quite trendy in mental health over the past few years, but it's in fact based on ideas within Buddhist traditions going back hundreds of years. The basic idea of mindfulness is a very simple one – it's the process of becoming aware of what is going on inside our minds, of stepping back and observing the way in which our minds work and becoming aware of our thoughts. Getting the hang of observing your thoughts without necessarily becoming caught up in them, or needing to react to them, can be very helpful in managing your mood. Remind yourself that:

- Thoughts are simply mental events taking place in our minds

- They are not facts

- They may or may not be true

- They may or may not be helpful to us

- We can choose which thoughts we hold on to and act upon and which we simply let go.

When people meditate they practice this technique of mindfulness – letting go of thoughts and simply watching them from a distance. The resources section in Chapter 10 tells you where you can find out about mindfulness.

TACKLE UNDERLYING NEGATIVE BELIEFS

Earlier we looked at the way in which negative beliefs about ourselves, the world and other people can make us vulnerable to mental health problems such as depression or anxiety. Sometimes even when we recover from these difficulties, these underlying negative views still remain, just waiting to be re-activated when we face difficulties in our lives, so they can trip us up all over again. When you have recovered from a mental health problem and are feeling well, this can be an excellent time to work on altering some of these negative ways of viewing the world. Low self-esteem or lack of confidence can make us much more vulnerable to mood problems. Spend some time thinking about your beliefs about yourself. Then start to plan how you can test out those beliefs – perhaps using the behavioural experiment ideas discussed in Chapter 3. Allow yourself to recognize the good things about both you and your world. The more you build up a positive view of yourself, other people and the world in general, the less weight those negative beliefs will carry, reducing their chances of damaging your mood.

LOOK AFTER YOUR RELATIONSHIPS

We all lead busy lives. It can be very difficult to prioritize our relationships with other people. However, research shows such relationships are really important in improving everyone's mental health. If we try getting close to others only when we need them, we are likely to find that they're unavailable. Spend time on a regular basis with the people who are important to you.

GET THE BALANCE RIGHT

Frequently, people with problems have very little balance in their lives. Think of the over-worked executive doing a 60-hour week, seldom seeing his family; the plumber so focused on keeping his business afloat that he never has time to slow down, relaxing with friends. Then there is the office-bound employee with no time for physical exercise or the exhausted mother with 3 small children who hasn't had an adult conversation for what feels like months. Though these are stereotypes, nonetheless they're often the kind of people who end up at their doctor's surgery with mental health problems. Take a look at your own life and daily routines. What's important for you to build in if you are to keep yourself as mentally well as possible? What needs to change? Small changes which keep your life in balance can make all the difference to your mental health.

So, what could Agata from our case study earlier have done differently following her period of depression to have reduced the chances of it happening again? Agata's done all the right things to help herself get over her depression. She's worked on her thinking patterns, become less self-critical and started including things in her routine to lift her mood. She's tested out some of her negative beliefs and gathered evidence that questioned her depressed view of herself and which helped her both feel and act in more positive, helpful ways.

However, she didn't recognize how important it would be to keep these things going both when she was well and, particularly, during stressful times. Agata also didn't deal with her underlying low self-esteem, leaving her still vulnerable to depression. Another factor is that she and her husband hadn't discussed how to work together to prevent a relapse in her mood. Finally, she hadn't really altered her view of her role in the relationship, not recognizing how important it is for her to feel good about her role as wife and mother, even though this doesn't include increasing the family's income. She certainly hasn't taken onboard how much she does which otherwise the family would have to pay someone else to do.

The good news? It's never too late. After this second period of depression, Agata learns from her relapse. She ensures she devises structures that will stand firm in both good and stressful times, and makes an agreement with her husband about this. She also works hard developing her self-esteem. Agata learns to recognize the good things she's doing, begins to appreciate her contribution and spends time developing skills which she knows she does have, which she can feel good about. The result? Agata feels better about herself, and is better equipped to manage when the inevitable stresses of family life come along. Most important of all, Agata gains strength from reminding herself of a phrase that helps her get through difficult times. Remember, if you too find yourself reexperiencing problems you thought you had completely overcome, then, like Agata, it's worth reminding yourself of that same phrase:

This, too, will pass.

CHAPTER 9

Change! Why you don't and how you can

The first step toward change is awareness. The second step is acceptance.

Nathaniel Branden

Have you ever asked for advice, and then replied to that advice by saying: 'Yes, but ...' Think about it for a moment. Aren't you really saying: 'No! Because ...' Many of us know what we should be doing about a problem. But actually doing it is quite another matter. We can feel really stuck, and be amazingly creative as to why NOT to do what we know we should. This chapter looks at why we don't always make those changes – and how to change that!

Are any of the below familiar reasons for not making change?

- 'If I stop worrying about my teenage daughter, bad things will happen to her.'

- 'If I try, I'll fail. Others will see how useless I am. And so will I.'

- 'If I'm weak, she'll take advantage of me.'

- 'It's the one pleasure I have left in life. I can't give it up.'

- 'There's no way you're going to get me to step onto an aeroplane.'

- 'I can't even get out of bed in the morning, and you're asking me to do homework?'

- 'It's him who needs to change, not me.'

- 'I hate exercise and I love chocolate; I can't change.'

In response to these, some therapists might say: 'Surely these clients just need to trust me and get on with therapy?' Or: 'I'll try and persuade them. It's just a matter of getting them to feel the uncomfortable feelings and do the work anyway, isn't it?'

If only it were that simple!

Dr David Burns is Emeritus Professor at Stanford University Medical School. He has been working with CBT for about 50 years, including alongside Aaron Beck, often called the founder of CBT.

Early in his career, David was curious why a large proportion of CBT clients didn't get better, despite using many different techniques, including those already described in this book. In his search for the answer, he developed the concepts of 'outcome resistance' and 'process resistance'. These concepts became part of a structure for therapy he called T.E.A.M.-CBT.[*]

T.E.A.M.-CBT stands for Testing, Empathy, Assessment of resistance and Methods, the key components identified to maximise successful therapy. Within this structure, outcome resistance refers to the *good* reasons why you may not actually want the outcome you're looking for, such as getting rid of anxiety or depression, or overcoming a habit or addiction, or developing a

---

[*] See reference at the end of the book.

closer, more loving relationship. Process resistance refers to the *good* reasons for not doing what's necessary to achieve the outcome you're looking for, such as doing work between sessions, facing your worst fears, giving up blaming the other person, or depriving yourself of something you really enjoy. These two types of resistance are the reason why people often say 'yes, but' and then go on to tell you why they actually won't be changing.

case study

Angela worries excessively about her teenage daughter. One advantage of her worry and anxiety is that it helps her to make sure that her daughter is as safe as possible. It also shows how much she cares about her. Her therapist could show her all the techniques in the world, but is she going to give up her anxiety when it protects her cherished daughter? That would be asking a lot. So, is she stuck with her crippling anxiety?

Outcome resistance

Probably. Angela's stuck, because she's experiencing something that can be called 'outcome resistance'. Does she really want the outcome for which she's gone to therapy (to get rid of her anxiety), when there are honestly very good and admirable reasons for her to hold on to her anxiety? Why would she want to stop worrying about her daughter, if it helps to keep her daughter safe? If she stops worrying, wouldn't it mean she also needs to stop caring so much?

But the worrying is draining and causes much frustration and misery. It's probably also badly affecting her relationship with her daughter. It also may be interfering with her daughter's need to take responsibility and be accountable. She appears to be stuck between a rock and a hard place!

So how can she become unstuck? Let's say Angela goes to a T.E.A.M.-CBT therapist for help with her crippling and exhausting anxiety. The therapist, at the right point in therapy, will probably ask: 'If I had a magic button right here, that you could press, and all your anxiety would be gone in an instant with no effort from you, so you'd be feeling relieved, energised, anxiety-free and confident, would you press it?' In response, Angela would probably say, 'Absolutely, yes!'

The therapist would then tell her that that's exactly what most people would say. They would continue with the offer that they have many techniques to do just that. But before going any further, the therapist would check if there were any good reasons why Angela might **not** want to do that, to press the 'magic button'. What are the benefits of her anxiety, how is it serving her or the ones she loves? Also, what does the anxiety say about her and her values that is truly beautiful?

Below are some possible answers Angela may give. Can you think of any more?

How does Angela's anxiety serve her? What benefits does it provide?	What does the anxiety say about Angela and her values that is beautiful and admirable?
It helps to keep her daughter safe by spotting all possible dangers.	It shows she really cares about her daughter's wellbeing.
It can help her to feel less worried, once she's thought about everything that could go wrong.	It shows she wants to be the best mum she can be.
It keeps her vigilant for any danger signals.	

Let's think about answers to similar questions for someone with a different problem.

case study

Jim is depressed. He is getting by in a job as a psychological therapist. However, despite many successes, there are several people he struggles to help and who aren't improving. Some even seem to be getting worse despite his considerable efforts. Although people try to tell him he's doing a really good job overall, he still feels low and keeps believing he's failing as a therapist. He's struggling to do his work and has gone off sick with depression. Furthermore, although he's been managing financially, he feels he should be doing much better, more like his brother who's earning twice as much working for a bank. On top of that, his wife's unhappy that he's become irritable and withdrawn at home.

If his therapist asked Jim to imagine a magic button that would instantly get rid of his low mood, such that he'd be happy and more confident in his work, less grumpy at home and relieved of financial worries, do you think he'd want to press it? Very probably.

However, what might be good reasons for Jim to keep hold of his depression? Jot down some ideas in the table below:

How does Jim's depression serve him? What benefits does it provide?	What does the depression say about Jim and his values that is beautiful and admirable?

Have you completed the table above? Great if you did, but for any readers who didn't, please have a go before you read the suggestions below.

How about these possibilities?

Maybe the depression gives Jim a break so he can conserve energy and recover.

Maybe it allows him to give up and stop banging his head against a brick wall.

And for the second column:

Maybe Jim's unhappiness means that his self-judgement is correct and he's realistic about not being an excellent therapist. Perhaps he feels he doesn't deserve to be happy unless and until he becomes an outstanding therapist.

Maybe it shows that he values being an excellent provider for his family.

If **you** are suffering from depression or an anxiety disorder, can you think of any benefits of your uncomfortable feelings and any great things they say about you and your values?

What are some **good** reasons to keep the negative feelings?

try it now
If you are suffering from anxiety:

How is your anxiety serving you? How is it helping?	What does your anxiety say about you or your values that is beautiful or admirable?

Let's do the same exercise for depression. If you're suffering from depression, and perhaps anxiety:

How is your depression serving you? How might it be helping?	What does your depression say about you and your values that is beautiful or admirable?

These questions might be hard at first. They're just not how you're used to thinking. The answers have probably been lurking just outside your conscious awareness. A therapist might be able to start you off with one or two suggestions:

Anxiety benefits

Does it protect you or your loved ones in any way?

Does it motivate you to remain vigilant?

Does it drive you to maintain high standards?

Anxiety positive qualities/values

Does it show how much you care?

Does it show that you value high standards?

Does it show that you care about what others think of you?

Depression benefits

Does it give you a break to lick your wounds?

Does it protect you from further negative judgement by encouraging you to lay low?

Does it protect you from futile effort in a hopeless cause?

Depression positive qualities/values

Does sadness show that you're human?

Isn't sadness often appropriate given what you might be lacking or have lost?

Does it show you're being realistic about the state of the world/others/yourself?

Does it show that you have high standards (even though you're falling short of them)?

We could repeat the exercise for all the other negative emotions such as feeling guilty, ashamed, inferior, worthless or incompetent. Lonely, unloved or abandoned. Embarrassed or self-conscious. Hopeless, pessimistic or despairing. Frustrated or angry.

case study

Back to Angela. The next question for Angela is: 'why *would* you want to give up all those benefits and compromise those beautiful qualities and values?' What's the value in getting rid of the anxiety?

It's actually a bit easier to answer these questions. How about:

'The negative feelings are just so horrible and draining.'

'They're really interfering with my life and my relationships.'

Angela's now likely to argue more strongly for the benefits of change, with no 'ifs', or 'yes buts', because they have been brought out into the open and the therapist is actually telling her why NOT to change, giving her all the 'ifs' and 'buts' she would otherwise be saying herself.

So now we've got arguments for Angela to get rid of her anxiety and arguments for her to hold on to it ... are we stuck? Well, suppose that instead of a magic button, we imagine a magic dial.

'You said you were feeling 100 per cent anxious. How much anxiety do you want, in order to still hold on to those benefits and those beautiful positive qualities and values? How much would be useful and healthy?'

The answer for Angela (and maybe for you) may well be somewhere between 5 per cent and 50 per cent, perhaps 20 per cent.

We've now made a fair and clear deal with Angela's subconscious, so that she can relax and trust the process. She may also have started to feel a bit better about herself, because she might have begun to see her anxiety differently. Rather than being a problem or an illness, she may be able to get a sense that the anxiety is coming from the very best side of her. It's showing what's right with her, rather than what's wrong with her.

She may see that she values the anxiety and wants to hold on to all the positive healthy and useful aspects of it, yet do some work to reduce its intensity so that it doesn't feel so uncomfortable and it doesn't continue interfering with her life.

Angela's now ready to make a clear decision and be highly motivated to dial down her anxiety.

If not, that's fine. It's up to her. The level of anxiety may be working for her, but now she may also be able to see it in a more accepting or even joyful way.

Before we move on and use methods to deal with Angela's anxiety or Jim's depression, there's something else we need to consider, and that's what Dr Burns calls Process Resistance.

PROCESS RESISTANCE

Having decided on a desired outcome, i.e. reducing anxiety to a certain percentage of what it has been, an individual needs to decide if they are willing to do the necessary work. In other words are they willing to take an active part in the process of change?

It means that some work is going to have to be done between sessions in order to get better. Constantly talking about the problems and having a therapist understand and empathise with them isn't going to be enough. As a general rule, clients who do 'homework' tend to get better, while those who don't, generally don't improve and may even get worse. There's also loads of evidence that with anxiety, facing fears rather than doing everything possible to avoid them is going to be a necessary part of the work. That's probably the last thing a person with anxiety is going to want to do!

Clients may think it's a good idea to just make a start on the therapy and hope that they'll somehow manage to do enough. Not a good choice! Clinical experience shows that it's essential to address the issue of whether the person's prepared to do the work right at the beginning of therapy. Are they willing to do the necessary work to gain the valuable rewards that they're hoping for?

The answer might be no. That's absolutely fine. It makes total sense. The therapist might be disappointed, because they'd love to see the client earn the rewards. But it's not *the therapist's* preference that's important. It's the client who's making the decision and they have to be accountable for that.

The T.E.A.M.-CBT therapist would stress the downsides and the size of the challenge, even to the point of trying to talk the client out of it. The client would then have to convince them that it's worth it and it's something she's willing and committed to do.

What is Angela's Process Resistance all about? Well, she is a busy working mum with no time to spare. So why would she choose to do this work (usually at least twenty minutes five times a week)?

It may well be that Angela is very motivated to put in the time and effort. However, to be totally successful some of her work needs to include facing her worst fears, making her anxiety temporarily even worse. It sounds too hard, or even impossible! She'll need to face her anxiety and do things like not checking up on her daughter, or sitting around worrying about her. Is Angela going to want to do that? Highly unlikely.

Is she willing to do that to dial down her anxiety?

It's up to Angela. If she wants all the benefits of successful treatment, it's a requirement. It's good to be clear in advance.

This approach has two benefits:

- It saves the client from wasting time and money on therapy that may be ineffective and take a long time.

- It's motivating in that the client becomes clearer about the good reasons *not* to do the work, and they can see that the therapist understands. This then frees them up to speak for all the benefits of doing the work.

Having done that, it's completely up to the client whether they want to continue to work with the therapist. No pressure! Alternative routes can be offered. Surprisingly, having seen all the good reasons not to do the necessary work, the process above usually results in clients being more motivated, with very few opting not to do the work.

The therapist doesn't have to argue for doing the work (that's become the client's job). The client would feel much less inclined to fight back when the therapist asks a lot of them.

case study

What about Jim? If he's not going to do 'homework' between sessions, he's unlikely to gain the potentially considerable benefit.

He's going to have to do the activity exercises and the written work to challenge his negative thoughts. Since he's suffering from lack of motivation, he's almost certainly not going to feel like doing these things. Is he willing to do what's necessary? It's up to him. No one else can do it. He's accountable.

It's useful to be clear about this right from the beginning. If Jim decides not to commit to doing the necessary work, he can always explore other ways of solving his problems. If that was Jim's choice, the therapist would respect that, but tell Jim that they would be very happy to work with him at another time, should he change his mind in the future.

try it now

Suppose we could make your goal of dealing with your problems come true.

Maybe that goal is to reduce your anxiety or depression. Increase your confidence and happiness. Have you jumping out of bed in the morning, looking forward to the day. Help you speak up at those meetings or give a confident presentation. Help you enjoy flying in an aeroplane. Help you stop procrastinating. Show you how to become closer to someone in your life.

Why not have a go using the exercises in this chapter?

However ...

The bottom line is that you would need to do regular 'homework'. If you're suffering from anxiety, the work will almost certainly involve facing your worst fears. If you're suffering from depression, you'll need to do activities despite initially not feeling like doing anything at all.

You may have many good reasons to not want to do the necessary work. It may not seem worth it to you. However, if you don't do the work, you're not going to get the benefits. It's your choice.

If your answer is 'no', then that's absolutely fine. No way would we try to persuade you.

If your answer is 'yes' ... are you sure? Why would you want to do homework when life is already difficult enough as it is? If you're suffering from a form of anxiety, why would you want to face your fears when you're already suffering from too much anxiety?

If you're still keen to go ahead, put your answer here:

Why you SHOULD have a go:

If you've convinced yourself that you are committed to doing the necessary work, then you can now move on to try the methods included in this book. The ones in Chapters 3 and 6 could be particularly helpful for anyone with anxiety and depression. No doubt some exercises will help, and some won't. Try not to be put off by methods that haven't helped you, and simply move on to finding those that do.

A NOTE OF CAUTION FOR THERAPISTS AND OTHERS WANTING TO HELP SOMEONE ELSE

Like the authors, you really want to help. It can feel so good to see someone overcome their problems and discover happiness and contentment. There may be an urge to push your solutions a little or to cajole, for the very best of reasons.

Dr Burns suggests that the most common cause for 'therapeutic failure' is trying to help the client. That may well appear odd and even paradoxical, so let's explain.

There's nothing wrong with trying to help. However, pushing, persuading, explaining or carefully manipulating is unlikely to be the best strategy.

Imagine if someone started gently but purposefully physically pushing you. In order not to fall over what would have to happen? You'd have to push back. It's a law of nature. Likewise, if anyone tries to push you to go in a certain psychological direction, you're going to have to resist, to maintain your equilibrium. It's a law of human nature.

In practice, that means that although you can offer to point out a path that can be helpful, it's up to the client or friend to decide if they want to walk it. No pushing or persuading. If there's even a whiff of manipulation, you will likely meet resistance. In practice that means 'sitting with open hands' and overcoming any urge to 'help'. That can be hard when you're witnessing a client or loved one suffering. It's going to require a lot of awareness and practice on your part to overcome old habits.

However, the rewards can be great. It means that as a therapist or friend, you can end up with someone who is really keen, accountable, and not 'yes but'ing you. It can also mean that you won't have to waste your energy working harder than your client or loved one, and all to no avail.

So we aren't going to try and persuade you in any way and will leave the choice with you. If you decide not to proceed, we honestly and deeply respect that, as there are probably plenty of good reasons for you not to take it any further. But for those of you who do want to go ahead, let's look at the following:

What sort of problem do you have? Anxiety, depression or improving a relationship? Or maybe it's a habit or addiction like nail biting, procrastination, overeating, gambling or drinking too much alcohol. Perhaps it's all of them! Maybe it's purely healthy grief, which is normal and only a problem if it becomes prolonged.

It's worth trying to work out what sort of problem you'd like to work on, or at least from which direction you'll come to tackle your group of problems.

That's because the types of resistance, the 'yes but's will differ between each group. You may also want to target different methods after dealing with the resistance.

Earlier in this chapter, we explored the meaning of outcome resistance and process resistance. Let's look at how they apply to a common problem:

HABITS AND ADDICTIONS

What are all the very good reasons to hold on to your habit or addiction and not make the effort to change?

case study

Sarah's addicted to chocolate. Getting in from work, she devours a whole bar of milk chocolate.

She dearly wants to stop, as she's aware of being overweight and the health implications that come with that. Sarah's tried various methods, but her considerable efforts never seem to last.

What are some benefits of Sarah's habit/addiction?

It's one of the few things that give her pleasure in life and which she has control over.
It just tastes so good.
She deserves it after a hard day's work!
Can you think of any more?

What are the disadvantages of her trying to change or give up this habit?

It will mean she has to deprive herself of the chocolate she so loves.
She'll have to tolerate the feelings that chocolate helps to comfort.
Can you think of any more?

What does this habit/addiction say about Sarah and her values that's positive and admirable.

It shows that she is not going to be dictated to by the opinions or advice of others.
That she values getting enjoyment out of life.
Can you think of any more?

try it now

If you have a habit or addiction you'd like to address, try to answer these three questions for yourself.

Take your time and try to think of as many answers as you can.

- What are some benefits of your habit/addiction?

- What are the disadvantages of trying to change or give up this habit?

- What does this habit/addiction say about you and your values that's positive and admirable?

If you've done the exercise thoroughly and given it enough time, you might have already benefited. Indeed, it may be all you need, because you already know what to do.

If you haven't done the exercise, then that too makes sense. You probably want to get on and read the rest of the book. Just remember you can't expect to derive the benefit either. Absolutely your choice.

However, if you're suffering from an addiction which requires detoxification, it's probably best to seek professional help, as self-help may well not be enough.

Right! If you've done the above exercise of listing the benefits of your habit/addiction, the disadvantages of trying to change, and what the habit/addiction says about you that's positive and valuable, you are probably feeling more motivated to start tackling it. That's because you've brought your partly subconscious resistance to conscious awareness. This is a powerful step in increasing motivation and reducing resistance.

RELATIONSHIP PROBLEMS

Maybe the best way to address the problems you're having is through addressing a relationship issue. This can be the most challenging area of all in which to deal with motivation and resistance.

It's about what a single individual can do to improve a relationship, rather than undertaking work as a couple.

case study

Marie is struggling in her relationship with her husband. He's spending too much time on his computer and doesn't contribute to family life. He seems distant and won't talk about his feelings. She's tried everything to get him to open up, but he just digs his heels in and becomes more distant. She's worried about the effects on her family and longs for him to show some warmth. She desperately wants a warm, loving relationship, but feels like it's hopeless wishing he'll ever change.

Outcome resistance

Let's spend a little time here. Marie may say she wants a more loving relationship with her partner. Are we sure that's actually what she's after? Perhaps she's asking for something slightly different? Maybe she simply wants her partner to be different, to behave differently, and she's not interested in making any changes herself.

Maybe Marie wants to find solutions to handling an abusive relationship. Maybe she just wants to complain about how hard things are.

You may say you want a more loving relationship with your partner, or a closer relationship with your friend. Are you sure? Could it be you're asking for something slightly different?

Maybe you just want your partner (or friend or co-worker) to be different, to behave differently.
Maybe you want to find solutions to handling an abusive relationship.

It boils down to three basic options:

Do you really want to become closer to this other person?

Do you want to maintain the status quo?

Do you actually want to distance yourself from them? Before you decide, it may be useful to consider process resistance for relationship work:

PROCESS RESISTANCE

To do this type of work, you're going to have to give up focusing on blaming the other person. That means focusing entirely on changing yourself and not trying to change the other person. Which makes sense, because in the end, you only have control over what *you* do.

Yes but …

I'm right and he's wrong.

It is *all her fault.*

I don't want to look weak.

Why should I always have to do all the work?

This type of resistance is perhaps the hardest to deal with, which is why it's so important to be truly committed to trying to get closer to the other person. Without that, it's unlikely that you'll do this challenging work. If you decide you don't really want to be closer to this person, that's fine. There are probably very good reasons for that. For instance, if your partner's a tyrant and truly abusive, it would make total sense.

Having said all that, you have enormous power to change the dynamic in a relationship by changing yourself, even if the other person is unwilling to change.

OK, so you're committed to working on becoming closer to the other person. Great! Before you decide whether you'd be willing to give up focusing on blaming them, let's just check a few things out.

THREE CRITICAL AREAS TO EXPLORE:
You may be thinking that:

1. There are advantages/reasons to focus on the other person's errors, e.g. 'Relationships should be 50/50. It's their turn to put in effort' etc.

2. There are costs of solely working on yourself, for instance 'I've already put in 150 per cent'.

3. The above two points show something about you and your values. For instance 'it shows I value truth, fairness, and mutually beneficial relationships'.

1. List the advantages and/or reasons to focus on the other person's errors and flaws:

2. List all the costs or disadvantages of solely working on yourself:

3. What do the answers to questions 1 and 2 above say about you that is positive and admirable? How are these connected to your values?

How did you do?

These are all *good* reasons *not* to give up focusing on the other person's errors and also why **not** to just focus on your own role in this dynamic.

Here are some good answers to the questions 1, 2 and 3 above:

1. *He's the one who needs to change.*
   It would be good for her to learn the error of her ways.
   I've already tried to be accepting and understanding and it hasn't worked.
   It feels good to put the blame on her.
   He needs help.
   She's toxic to the children and needs to change for their sake.

2. *It will be hard work.*
   It will be uncomfortable looking at any faults that I have.
   I'll be vulnerable.
   I'm already exhausted.
   I'll need to put up with her.

3. *I value a good family life.*
   I really care about the children.
   I don't believe in being a doormat, I care about my dignity.
   I believe that people should be thoughtful and kind.
   I really want to help my partner.

So, here's the next question:

> **Considering all these powerful advantages of focusing on the other person's errors and the cost of working on yourself, why would you want to move forward with this hard work? Have a go at writing down some answers before reading our suggestions below.**

Here are some possible answers:

I've tried many other ways.

I really want this relationship to work and I'm willing to do the work.

I won't know what's possible unless I try.

I could really benefit from some self-development.

I know that there are many aspects of my behaviour and attitude that I could work on, and I'd like to find out how.

I feel hopeful and pleased that my therapist is being honest about what's required.

It may be that you're not able to complete the tasks above just now, and that's fine. There are many reasons that may be the case, and we are not going to force you! However, if you don't do the exercises, it's unlikely you will gain any benefit.

So, how willing are you to change? From 0–100?

Once you're sure you really do want to be closer to the other person, whether it be partner, friend, work colleague or someone else, and that you're willing to give up blame and focus entirely on what you can do differently, the next step is to learn and practise the skills. With commitment and determination, it really is possible to master the steps and transform your relationship. However, there will always be motivational and resistance issues involved in putting the steps into practice.

REMEMBER

By now, you should have a good understanding of the critical role of motivation and resistance. If you feel reluctant and lacking energy to tackle the personal work you're contemplating (and that's almost always the case at some level) then this needs to be dealt with. Without such work, whatever methods are chosen, they're almost certainly going to be less effective or not work at all. If the work is done, lots of CBT methods, some of which are described in this book, will be effective, some more than others.

You can find out much more on the T.E.A.M. approach by looking at the work of Dr David Burns. Please see the appendix at the back of this book for more information.

So, to sum it all up, 'People don't resist change. They resist being changed!' — Peter Senge.

And now it's over to you. You have a whole new toolkit for what to do when you are stuck. Think of the work you can do on yourself, and the satisfaction and peace it could bring.

try it now
PROBLEM 1
How is it helping you?

What good things does it say about you?

What bad things would result if you got rid of the problem entirely?

How would life be better without the problem?

And with these answers, you can now decide whether this is indeed the time to change, and use this as motivation to make it happen. Explore and enjoy using the techniques in the other chapters of this book to help you get rid of as much of the problem as you wish, but to keep the good bits which you so value about yourself.

Now let's look at another difficult problem which affects a huge number of people.

T.E.A.M.-CBT FOR CHRONIC PAIN

Being constantly in pain is often mentally and emotionally upsetting. Long-term or chronic pain can be difficult to treat. Many with chronic pain have tried to get rid of the pain using various physical treatments, like pain-killing tablets, physiotherapy, injections, perhaps acupuncture, and in more extreme cases even surgery.

Unfortunately, these seldom work in the longer term and people can be left with severe pain, plus the feeling of being mentally and physically upset. It can mean a great deal of suffering, and can severely disrupt many areas of your life, including work, mood, self-esteem and your relationships with other people. The good news is that CBT approaches can help deal with the suffering.

Using T.E.A.M.-CBT can help enormously in dealing with a big part of suffering, namely the mental and emotional distress you feel when you are living with chronic pain. Many people with chronic pain frequently suffer with feelings of intense frustration, anger, guilt, anxiety, worry, depression and shame.

In many cases the more upset you get, the worse the pain gets. Then, the more severe the pain, the more upset you become. It's a vicious circle, where one feeds the other (see Diagram 1). Some emotions, for example anxiety, also lead to physical sensations such as bodily tension, which can then increase the pain. Likewise, people often try to 'push through the pain', leading to another vicious cycle of pain flare-ups, then needing a long rest, and then more physical tension (see Diagram 1).

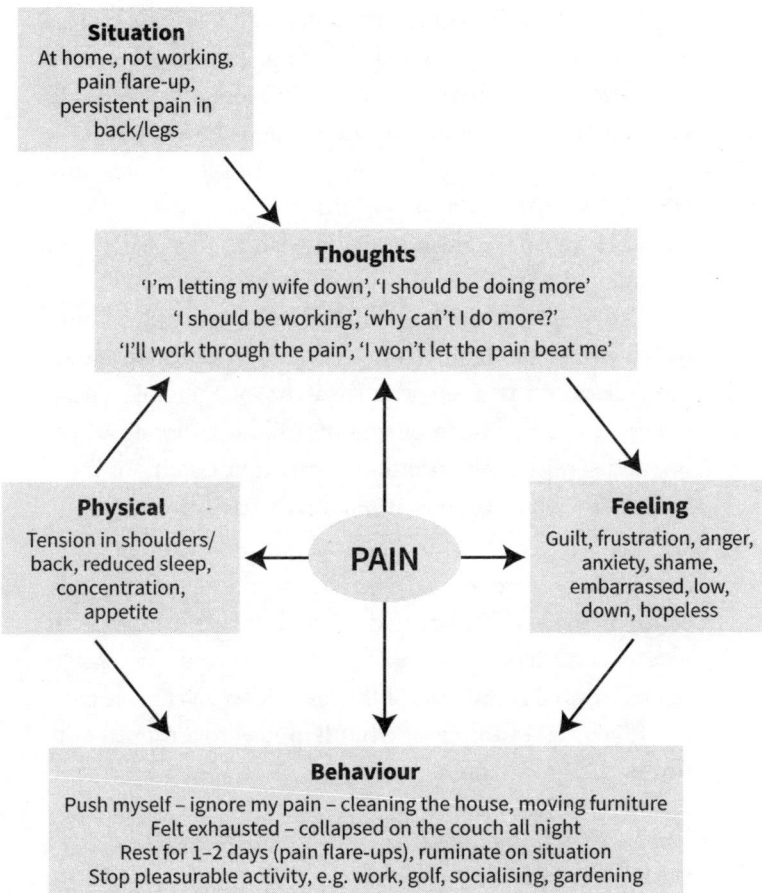

Diagram 1 – CBT formulation of chronic pain

The experience of chronic pain is often complicated, involving biology, psychology and life circumstances, so chronic pain is often described as a bio-psycho-social phenomenon.

This may all sound rather interesting, but what can we actually do about it, and where do we start?

How about learning to change how we feel? This could reduce our emotional suffering, and maybe enable us to also learn to cope with the pain more effectively. Do you reckon that could help you manage chronic pain?

WHERE TO START?

What if we offered you a magic wish to deal with your chronic pain, what would you wish for right now?

You probably said **GET RID OF** the pain. That's totally understandable. Nobody wants to suffer unrelenting pain and if you've had it for years, or even decades, it will have caused enormous mental and emotional suffering and significantly interfered with your quality of life.

Let's look at what you've already tried.

Pain journey exercise

Many people will have tried a variety of pain treatments before considering psychological treatments. We highlighted several at the start of the discussion on chronic pain.

Here's an example of what happens to a typical person with chronic pain.

Goal – Eliminate the pain

Table 1: Pain journey exercise

Treatments Tried	What was the outcome of the treatment in the short term?	What was the outcome of the treatment in relation to your goal?
Medication	Reduced pain for 2–3 hours. Had side-effects.	Goal not achieved. Pain still there.
Physiotherapy	Helped increase range of movement. Pain slightly worse.	Goal not achieved. Pain still there.

Treatments Tried	What was the outcome of the treatment in the short term?	What was the outcome of the treatment in relation to your goal?
TENS Machine	Eased pain while using TENS machine.	Pain still there when machine removed.
Acupuncture	Eased pain for several weeks.	Pain returned and persists.

So now it's your turn. If you do suffer from chronic pain, which is presumably why you are reading this section, please list all the treatments you've previously tried to get rid of your pain, writing them in the left-hand column of the table below.

Goal – Let's use the same one of the typical person – Eliminate the pain

Table 2: Pain journey exercise

Treatments Tried	What was the outcome of the treatment in the short term?	What was the outcome of the treatment in relation to your goal?

Once you've written this down, take a moment to review the list. Then ask yourself this key question:

What conclusions are you coming to about your goal, given these treatment results?

This may be hard to answer and accept. You may be feeling more hopeless and despairing about the outcome now and wondering whether your goal is realistic, given your experiences. That's OK. It's normal and understandable to feel this way as you think back on the things you've tried. There've often been many cycles of hope at starting a new treatment, followed, unfortunately, by disappointment afterwards at the failure to eliminate the pain.

Diagram 1 (page 191) shows how chronic pain is a complex experience, and that being in persistent pain often leads to considerable mental and emotional distress.

The table below shows a list of common emotions people struggle with when they experience chronic pain. Tick the ones you've had, plus any of the thoughts that might go through your mind, adding ones not yet mentioned into the box marked 'other'.

Which of these emotions and thoughts do you struggle with when you're in pain?

Emotional Suffering

Emotion	Thoughts (examples)
Frustration	*I should be able to do things quicker* *I'm only 40 but feel 70*
Anger	*I shouldn't be like this* *It's unfair being in pain all the time*

Emotion	Thoughts (examples)
Guilt	*I'm letting my family down* *I should not have to lean on other people*
Shame	*I shouldn't have to ask for help* *I can't work*
Anxiety	*What if things get worse?* *What am I going to do next?* *I can't cope*
Fear	*What if I end up in a wheelchair?*
Depression/ low/hopeless	*I feel useless* *I can't do what I used to do* *What's the point, I won't enjoy it* *There's nothing I can do*
Other	

Now here's that magic wish question:

If you could have a miracle to deal with the above **suffering**, what would you wish for?

If you choose to eliminate it, that's good! But guess what? You don't need a miracle to do this! T.E.A.M.-CBT can help you overcome your intense feelings of guilt, frustration, fear and anger, allowing you once again to feel more positive emotions such as pleasure, joy, peace, and move towards acceptance.

However, it might be a mistake to eliminate **all** our suffering. Let's take a closer look at why that's so.

OUTCOME RESISTANCE

In many cases people experience intense emotional suffering. Frustration, guilt and anger are some of the most common of these distressing feelings.

case study

Let's look at Michaela's problems. She has suffered with chronic back pain for ten years, has been diagnosed with fibromyalgia and experiences widespread pain and fatigue. She experiences a lot of emotional distress including guilt, frustration and anger:

Guilt: 8/10 intensity. Thoughts such as 'I'm letting my children down', 'I should be working'.

Frustration: 9/10 intensity. Thoughts such as 'I should be able to do the housework in a day', 'I'm 30, but I feel 55 most days', 'I can't pick up my son'.

Anger: 8/10 intensity. Thoughts such as 'Why me?' 'I should not be like this!', 'Why don't people understand/listen?'

Imagine there's that red button and if Michaela pressed this all her feelings of anger/frustration/guilt would vanish instantly, along with all those upsetting associated thoughts that have been distressing her. She would feel happy and relaxed with no effort at all. So, do you think she should press the button?

Probably by now you can guess **NO** (remember the button's only removing the emotional distress, not the reality of her persistent pain). If she did press it, it would be like saying, 'I want to feel happy and relaxed while experiencing persistent pain and not doing valued activity.'

So, let's think why it would be a mistake to press the red button.

The two key questions to ask are:

- What do her feelings and thoughts show that's really **valuable** and **important** about Michaela?

- What are the advantages and benefits of having these thoughts and feelings?

Put your answers to the questions into the values and advantages columns.

Positive reframing table

Thoughts & feelings	Values What do Michaela's thoughts and feelings show that's really valuable and important about her?	Advantages/ Benefits What are the advantages and benefits of Michaela having these thoughts?
Guilt *I'm letting my children down.* *I should be working.*		
Frustration *I should be able to do the housework in a day.* *I'm 30 but I feel 55 most days.* *I can't pick up my son.*		
Anger *Why me? I should not be like this!* *Why don't people understand/listen?*		

Hopefully you thought of some answers. Often this is really difficult to do at first, as we're so used to thinking of feelings of guilt, frustration, anger and other similar emotions as NEGATIVE feelings that have nothing good about them at all.

Here are some ideas of what it really shows about Michaela.

Positive reframing table

Thoughts & feelings	Values What do Michaela's thoughts and feelings show that's really valuable and important about her?	Advantages/Benefits What are the advantages and benefits of Michaela having these thoughts?
Guilt *I'm letting my children down. I should be working.*	She cares about her children and family.	Caring about her family helps her feel good and connected.
	She values work and contributing.	Working helps her feel useful as a person and feel she's contributing to her family.
	She values independence.	Being independent makes her feel good and in control of her life.
	She has high standards and expectations.	Her expectations motivate her to achieve things in life.
	Michaela values her conscience, sense of right and wrong.	Her sense of right and wrong is like a compass – it guides her what to do in life.
	She values her sense of responsibility.	Being responsible means she looks at herself and is accountable.

Thoughts & feelings	Values What do Michaela's thoughts and feelings show that's really valuable and important about her?	Advantages/Benefits What are the advantages and benefits of Michaela having these thoughts?
	She values being accountable. She values honesty.	Helps her be reliable for others.
Frustration *I should be able to do the housework in a day.* *I'm 30 but I feel 55 most days.* *I can't pick up my son.*	She values her role as housekeeper for the family. She values achievement. She has a sense of determination. She values a sense of responsibility. She values expectations. She values honesty and a sense of realism. She values passion/desire.	Her role gives her an identity and caring for the family makes her feel good. Achieving things makes her feel she's contributing. Being determined helps her keep going when she doesn't feel like it. Being responsible makes her reflect and shows her morals. Her expectations motivate her to work and achieve things. Honesty helps her face the reality of life. Her desire to support her family gives her a purpose.

Thoughts & feelings	Values What do Michaela's thoughts and feelings show that's really valuable and important about her?	Advantages/Benefits What are the advantages and benefits of Michaela having these thoughts?
Anger *Why me? I should not be like this!* *Why don't people understand/ listen?*	She has a conscience – a sense of right and wrong. She values fairness and justice. She cares about the opinion of others/ values support. She has high standards. She values respect. She stands up for herself. She has a sense of realism.	Shows she's got morals. Anger tells her if she feels something is wrong. Fairness and justice help guide her in life. Caring about others helps her listen to them, support from others is crucial in life. Anger can motivate her to take action or stand up to others if she feels she's being disrespected. Respect helps her value others and herself. Its normal to feel angry when pain interferes with your life constantly – she's human.

Now when you look at this list of valuable things, it shows a great deal about Michaela and the benefits of her distressing emotions. Would Michaela want to lose all of them?

The answer is probably no, because in T.E.A.M.-CBT we realise that a lot of our mental and emotional suffering comes from when we struggle to live up to our values, and our values are central to our identity. They give us drive and purpose and have many advantages in life.

try it now

Try the Positive reframing exercise for yourself. First, identify which distressing emotions you want to change and rate the intensity of each emotion (0–10 scale, where 0 = no distress, 10 = extreme distress).

Then write down the values and benefits of these emotions in the table below. Here we go ...

Positive reframing table

My thoughts and feelings	Values What do your thoughts and feelings show that's really valuable and important about you?	Advantages/ Benefits What are the advantages and benefits of you having these thoughts?
For example, guilt		
For example, frustration		
For example, anger		

Did you complete the Positive Reframing exercise? If so, you may well find just doing it has already started to change your view and perspective on these emotions and has reduced the intensity.

Would you still want to press the button and get rid of all these values and benefits? Probably not, and with good reason! But it doesn't have to be all or nothing. There is an alternative choice.

Magic dial

Imagine you have another choice. Instead of the magic button you could use a *magic dial*. This lets you choose what intensity of the emotions you can keep, so you still can hang onto all the values and benefits that are part of the package.

Three choices:

1. Turn up the dial – increase your suffering (if you want to).

2. Keep the dial the same – keep your suffering at the same level.

3. Turn the dial down – keep all the values and benefits associated with your emotions, but lose the unhealthy emotional intensity which causes you to suffer intensely.

By deciding what level you want to reduce your emotional intensity to, you accept that the emotions do have some value and benefits for you. And strangely enough, when you begin to accept something, it can be easier to change, as you realise there is a good/healthy side to each emotion and so you start resisting less and less.

So, what would you want to do with the dial for your emotions?

Write down your answer below ...

Sometimes Positive Reframing is like re-educating yourself about emotions. We've been conditioned to see emotions like anger, depression and guilt as generally negative, therefore we try to eliminate or avoid them. In T.E.A.M-CBT we believe there's a healthy amount to each emotion, which represents part of your value system and associated benefits.

PROCESS RESISTANCE

Now, if you've decided your goal is to reduce your own experience of emotional intensity, we are getting ready to bring about change.

The final element is how willing you are to put the effort in to achieve your goals. Often in life the theory and knowledge of change is relatively easy to understand. The learning how to do it and how to apply this knowledge is another matter entirely! It often requires great motivation to change, as change can be challenging and may mean facing up to emotions and thoughts you've been avoiding, and learning to let go of old patterns and habits of thinking that have existed for a long time, which feel like they're actually part your identity, basically who you are.

The requirements

To achieve change, you have to be willing to do the following:

- Being ready to *focus entirely on your role in pain management*, looking at your thoughts, feelings and behaviours, even when this feels extremely painful and difficult.

- *To do homework daily*, like keeping thought diaries and practising regularly, to learn the specific skills required to challenge thoughts, even when you don't feel like it.

In nearly all types of CBT, willingness to regularly do homework tasks is an essential ingredient in bringing about change.

Key question: *How willing are you to put this required effort into achieving your goals?*
(rate this on a **0–10** scale, where **0** = no motivation at all to change, up to **10** = highly motivated to change)

If you score between **8–10,** that shows you're highly motivated and the outlook for change is very promising, as the methods and techniques used in T.E.A.M.-CBT have been used successfully by many people and helped them achieve their goals. If this is you, you can move onto the next section, 'Methods'.

If you score less than **8**, there is part of you that will resist change and struggle with the vital process requirements.

What are all the good reasons not to change?
Some possible examples of resistance could be:

- They still might offer me an injection or tablet that could fix my pain.

- If I accept this, it's like giving in and being defeated.

- If I give in and accept this, I'll become even more depressed.

- I've got to keep working. I can't change and pack in my job.

- Changing my thinking is not going to change anything. The pain's still there.

If there are lots of good reasons *not to change* then that's fine. Maybe now is not the right time. Sometimes people need more time and aren't prepared to accept that they have to focus exclusively on their role in pain management, and stop searching for an external solution. You can always come back when you feel ready.

METHODS
We are now ready to look at how to change the emotional intensity and achieve the goals you set yourself at the end of the positive reframing exercise. Within T.E.A.M.-CBT there are over 100 methods and techniques which have all proved valid and useful at helping bring about change. Here are some examples of the methods that people have found helpful with chronic pain, once they have completed the previous steps.

One of the most useful techniques is:

The Acceptance Paradox

If you can accept a thought, you can usually reduce the emotional suffering that comes from struggling to accept reality. The idea of acceptance may have originated from Buddhism, where suffering is seen to be caused from our habit to pick and choose in life, e.g. 'I want this and I don't want that'. This picking and choosing can sometimes lead to mental and emotional suffering, especially when you do not get what you want/expected in life.

This often means letting go of our demands for reality to be different than it is. Acceptance is not easy, but you can learn this skill, and this may lead to more peace and a reduction in emotional suffering.

The key for any of these techniques to be effective is that they not only give us an intellectual answer but most importantly are acceptable on an emotional level. Example: *I should not be in pain*. Belief in the thought – 100 per cent.

Acceptance thought: *It would be ideal if I was not in pain*. Belief in this thought – 50 per cent.

In this case the person only believes the alternative thought 50 per cent. A big part of them isn't accepting it. Intellectually it may seem logical, but on the gut level we don't believe it. So, this won't work to overcome the original thought. The belief in a positive thought needs to be 100 per cent, or very close to it, for you to have a strong chance to truly feel an emotional change on the gut level. This is the challenge in T.E.A.M.-CBT; to develop 100 per cent positive thoughts that can overcome the original thought.

The good news is there are many techniques available to help us find what works for us. You may find some of the techniques work best with the help of a therapist at first, as they can demonstrate how to use the techniques effectively.

The semantic technique

Our emotional suffering often stems from the meaning we give to the words of our thoughts. If we can change the meaning, by changing the words in our thoughts, this can help reduce the emotional suffering. Often when we are feeling very frustrated, guilty and angry or upset, we may be using 'should statements'.

For example, sometimes when we're feeling frustrated with being in pain, we say to ourselves:

'I should be able to do more', 'I should not have to go so slowly'.

The word 'should' has a powerful meaning behind it. It is a very demanding word, especially if you're struggling to achieve a task due to the impact of chronic pain. When you tell yourself you should, you are probably being very self-critical and this can lead to more frustration, guilt and low mood.

This technique gives you the choice to substitute the word 'should' for a less critical, more realistic statement.

Here are some suggestions:

- It would be nice if …
- It would be wonderful if …
- It's OK if …
- It would be ideal if …
- I would prefer it if …

It's important to be realistic. You need to identify what's currently limiting you. Otherwise, these statements may not be convincing.

For example, saying to yourself 'it would be nice to do more' may not work for you and may feel a bit simplistic.

To overcome this, the technique needs to incorporate some elements of acceptance of our limits (i.e. being in pain) and the situation itself. This can be difficult but may lead to a more believable thought.

For example:

- 'It would be ideal if I could do the housework in one day like I used to do. But if I try, I suffer for three days with a pain flare-up and will end up in bed.' Belief in thought – 100 per cent.

- 'I would prefer to do the housework in one day, but my pain stops me. I can learn to accept my new limits and instead focus on what I can do.' Belief in this thought – 95 per cent.

Feel free to experiment and substitute your 'should' statements for less harsh words. You may find this helps and softens the emotions, reducing self-criticism.

The previous chapters in this book also give ways to manage your thoughts differently. There are also many other techniques in the book *Feeling Great* by David D. Burns, creator of T.E.A.M.-CBT.

So now it's up to you. Do you really want to change? Start thinking about what it is you would like to change about yourself – and then work your way through this chapter again. You could be absolutely delighted with the results!

SPECIAL THANKS FOR THIS CHAPTER GO TO THE FOLLOWING PEOPLE:

Dr Peter Spurrier who is a retired GP with over ten years of experience in CBT, coaching and mentoring. He is the founder of T.E.A.M.-CBT UK and has worked with T.E.A.M.-CBT for eight years.

Derek Reilly who is a CBT Psychotherapist of 21 years. He specialised in treating chronic pain for 20 years, within a multi-disciplinary Chronic Pain service. He has worked with T.E.A.M-CBT for eight years and is publishing an audit of clinical outcomes of T.E.A.M.-CBT in the treatment of chronic pain.

Dr David D. Burns is an American psychiatrist and adjunct professor emeritus in the Department of Psychiatry and Behavioural Sciences at the Stanford University School of Medicine. He is the author of several bestselling books and is the creator of T.E.A.M.-CBT. To learn more, visit his website Feelinggood.com

CHAPTER 10

What now?
Further helpful resources

We hope that you've found this book useful and interesting. However, there are many other resources out there which can assist you even further in overcoming difficulties or making the life changes you would like to make. Most people benefit from multiple sources of help. This chapter contains our recommended resources for accessing additional help if you are facing anxiety, depression or other emotional problems. We have also included resources for people facing relationship difficulties or managing physical health conditions or chronic pain. The resources range from people in your immediate circle, to professional and voluntary services, to additional self-help alternatives. Most of these resources tell you about other sources of CBT help, but we have included some that offer alternative types of help as well. Following on from this book, you may find it extremely helpful to explore the many other excellent books, websites, CDs and DVDs that are available.

WHAT'S RIGHT FOR YOU?

Take a quick browse through the range of resources that follow. In the following table, write down the ones you are going to try. You might want to add additional resources you have discovered to the table, and even make notes of what you find when you try them out.

Resources I am investigating

Additional resources I have discovered

My comments on the resources investigated

FAMILY AND FRIENDS

Almost everyone turns to friends or family for help from time to time, and we certainly encourage you to do so. Although your friends and family probably don't have the expertise to do more than support and listen to you, this can be invaluable at stressful times. There are plenty of organizations providing support and information to your family and friends too, if they want to support you but aren't sure how. Encourage them to find information from the sources suggested.

Avoid trying to make your loved ones feel responsible for your mental health. Doing so may jeopardize your relationship and is unlikely to help you in the end. You must take ownership of the task of dealing with your difficulties, getting better, maintaining improvements and coping with any relapse. Alternatively, do ask for professional help if you feel you can't do this by yourself.

FINDING PROFESSIONAL HELP

There are many ways to access professional help for your difficulties. Many of the types of professionals mentioned here will offer CBT; some may offer different types of psychological therapy. We would recommend that before you start work with a professional you ensure that they are trained in those treatments that have a sound evidence base and are recommended by the National

Institute of Clinical Excellence (NICE) for your condition. The NICE website at www.nice.org.uk will give you more information about this. It is always a good idea to inform yourself about the types of treatment you might find helpful. Don't be afraid to ask for what you think you need – but do also be open to changing your mind if you are offered advice that a different approach would be best for you.

THE NATIONAL HEALTH SERVICE (NHS)

The advice is this section is tailored to those within the UK who have access to the National Health Service (NHS). If this doesn't apply to you, some of the advice is likely to still be relevant, or you may want to jump ahead to the section on private healthcare.

There is always the possibility that your anxiety, depression or other emotional difficulties are due to physical health problems. Your GP is the person to rule this out and then refer you on to the most appropriate mental health professional if necessary. It's therefore normally a good idea to go to see your GP in the first instance.

Your GP is the first point of access to psychological services and will be able to assess your symptoms and advise you on services in your area. Services do vary from area to area in terms of the types of treatment provided, waiting times and the number of sessions which can be offered. Many GP surgeries have in-house counsellors who may work in a variety of ways, perhaps including CBT. Additionally, if you are over 18, in many areas of the UK your GP will have access to new services called IAPT services (IAPT stands for 'Improving Access to Psychological Therapies'). See www.iapt.nhs. uk for more information. All IAPT services will offer CBT as one choice of treatment. If you are particularly interested in CBT you should let your GP know this.

Some IAPT services will allow you to self-refer. You can usually find these by searching the internet for IAPT services in your area. They will often also offer appointments outside of normal working hours in order to make help more accessible for working people.

If you are under 18 your GP will be able to refer to sources of help specifically for young people and their families.

If your GP feels your difficulties are somewhat more severe he may recommend that you are referred to a Community Mental Health Team (CMHT). These teams are generally made up of psychiatrists, psychiatric nurses, clinical psychologists, occupational therapists and social workers. They may also have some mental health support workers. Any of these professionals might be involved in your care. The clinical psychologists within these teams will often be able to provide CBT.

Clinical psychologists may also work in primary care in some areas. They have a doctoral qualification, for which they've undertaken extensive training in the understanding and treatment of mental health problems. They primarily utilize psychological therapy, which involves one-to-one or group sessions where you work together on overcoming your difficulties. For the best results, make sure your psychologist is familiar with scientifically validated therapies and if you are specifically interested in CBT then ask about this. If you particularly want to see a clinical psychologist then do ask your GP about this.

A word about medication – your GP may feel that you should try medication for your condition. They may also ask you to see a psychiatrist so as to get a more expert opinion on what medication might be right for you. Psychiatrists are doctors who specialize in illnesses of the mind rather than the body. They have extensive training in diagnosing and treating mental disorders.

Most psychiatrists primarily use medication in treating these disorders and can help you to manage any side effects the medication may cause. Many people feel very worried about taking medication. However, we would recommend that you don't automatically rule this out. Medication can be very helpful to people and can sometimes work well alongside psychological therapies (although this may not be true in the case of anxiety – see

Chapter 3). We have talked about medication elsewhere in this book so do look at Chapter 6 for more information about this. Your doctor will be able to discuss all your questions with you. There is also a wealth of information available on the internet but do be selective about what you read. The Royal College of Psychiatrists has some useful information on medication on their website at www.rcpsych.ac.uk.

PRIVATE HEALTHCARE

If you have private health insurance, this may include cover for psychological therapy. You may also choose to pay for private therapy yourself. Once again, it is often best to go through your GP, but private medicine may offer a wider choice of specific therapists and of therapies, including complementary therapies. Insurance companies normally have a list of the practitioners they recognize. However, many therapists may not be on these lists but still be appropriately qualified. Various professional organizations have lists of qualified practitioners.

If you are specifically interested in finding a CBT therapist and are based in the UK, then we would recommend you first look at the British Association for Behavioural and Cognitive Psychotherapies website at www.babcp.com. Other organizations which will have lists of qualified private therapists offering different forms of psychological therapy (sometimes including CBT) are the British Psychological Society at www. bps.org.uk, the UK Council for Psychotherapy at www. psychotherapy. org.uk or the British Association for Counselling and Psychotherapy at www.bacp.co.uk.

We strongly recommend that you find out about the qualifications of anyone you would like to see before making an appointment. Each professional organization's website has a 'practitioner search' to help you. If someone is accredited by one of these professional bodies it means they have gained appropriate qualifications in their particular area of expertise. Any professional should be willing to discuss their qualifications and approach if you have questions for them.

SELF-HELP BOOKS
Bookshops and libraries display a bewildering array of self-help books. It can be very difficult to know where to start! It's important to be aware that quite a few of these books aren't based on scientifically validated treatments. This doesn't necessarily mean they won't be helpful for you as an individual, but you are probably more likely to feel confident using a book that has a strong and clear evidence base. Below is a list of the books which we and our patients have found most useful. These books give solid help based on well-researched strategies for alleviating emotional distress. Most are based on CBT principles but some offer alternative approaches which we believe can work really well alongside CBT

techniques. Some are quite new; some are older but still very relevant to people today. Obviously these aren't the only good books out there but these are the ones we most frequently recommend.

BOOKS BY THE AUTHORS

Overcoming Depression for Dummies by Elaine Iljon Foreman, Laura L. Smith and Charles H. Elliott (Wiley, 2008).

Overcoming Anxiety for Dummies by Elaine Iljon Foreman, Charles H. Elliott and Laura L. Smith (Wiley, 2008).

Anxiety and Depression Workbook for Dummies by Elaine Iljon Foreman, Charles H. Elliott and Laura L. Smith (Wiley, 2009).

Fly Away Fear: Overcoming Fear of Flying, by Elaine Iljon Foreman and Lucas Van Gerwen (Karnac, 2008).

BOOKS ON DEPRESSION

CBT

Choosing to Live: How to Defeat Suicide through Cognitive Therapy by Thomas E. Ellis and Cory F. Newman (New Harbinger Publications, 1996).

Cognitive Therapy of Depression by Aaron T. Beck, A. John Rush, Brian F. Shaw and Gary Emery (Guilford Press, 1987).

Overcoming Depression by Paul Gilbert (Robinson, 2000).

Overcoming Low Self-Esteem by Melanie Fennel (Robinson, 2009).

Overcoming Depression: A Five Areas Approach by Chris Williams (Arnold, 2001).

Overcoming Depression One Step at a Time by Michael
E. Addis and Christopher R. Martell (New Harbinger
Publications, 2004).

Alternative approaches
*The Mindful Way through Depression: Freeing Yourself
From Chronic Unhappiness* by Mark Williams,
John Teasdale, Zindel Segal and Jon Kabat-Zinn
(Guilford Press, 2007).

BOOKS ON ANXIETY, PHOBIAS, PANIC AND STRESS
CBT
Master Your Panic and Take Back Your Life! by Denise
Beckfield (Impact Publishers, 2000).
Mastery of Your Anxiety and Panic by David Barlow and
Michelle Craske (Oxford University
Press, 2007).
Mastery of Your Anxiety and Worry by Michelle Craske,
David Barlow and Tracy O'Leary (The Psychological
Corporation, 1992).
The Relaxation and Stress Reduction Workbook by Martha
Davis, Elizabeth Eshelman and Matthew McKay
(MJF Books, 1995).
Overcoming Anxiety: A Five Areas Approach by Chris
Williams (Hodder Arnold, 2003).
*Overcoming Anxiety Self-Help Course: A 3-Part
Programme Based on Cognitive Behavioural Techniques*
by Helen Kennerley (Robinson, 2006).
*Overcoming Panic and Agoraphobia Self-Help Course:
A Three Part Programme Based on Cognitive
Behavioural Techniques* by Derrick Slove and Vijaya
Manicavasagar (Robinson, 2006).

The Anxiety & Phobia Workbook by Edmund J. Bourne (New Harbinger Publications, 2005).

Alternative approaches

Embracing Uncertainty by Susan Jeffers (Hodder & Stoughton, 2003).

Feel the Fear and Do It Anyway by Susan Jeffers (Vermilion, 2007). This is a reprint and update of a book first published in 1987.

Learn to Relax: Proven Techniques for Reducing Stress, Tension and Anxiety – and Promoting Peak Performance by C. Eugene Walker (Wiley, 2000).

BOOKS ON SOCIAL ANXIETY OR SHYNESS

CBT

The Shyness and Social Anxiety Workbook: Proven Techniques for Overcoming Your Fears by Martin Antony and Richard Swinson (New Harbinger Publications, 2000).

Dying of Embarrassment: Help for Social Anxiety and Phobia by Cheryl Carmin, Alec Pollard, Teresa Flynn and Barbara Markway (New Harbinger Publications, 1992).

Overcoming Social Anxiety and Shyness by Gillian Butler (Robinson, 2009).

BOOKS FOR BETTER GENERAL MENTAL HEALTH

CBT

Mind Over Mood: Change How You Feel by Changing the Way You Think, Second Edition by Dennis Greenberger and Christine A. Padesky (Guildford Press, 2015).

The Feeling Good Handbook by David D. Burns (Plume, 2000).

Reinventing your Life by Jeffrey E. Young (Penguin, 1998).

Manage your Mind by Gillian Butler and Tony Hope (Oxford University Press, 1995).

Mindfulness: A Practical Guide to Finding Peace in a Frantic World by Mark Williams and Danny Penman (Piatkus, 2011).

Other Approaches

Acceptance and Commitment: An Experiential Approach to Behaviour Change by Stephen Hayes, Kirk Strosahl and Kelly Wilson (Guildford Press, 1999).

Authentic Happiness: Using the New Positive Psychology to Realize Your Potential for Lasting Fulfillment by Martin E. P. Seligman (Free Press, 2004).

Feeling Better, Getting Better, Staying Better: Profound Self-Help Therapy for Your Emotions by Albert Ellis (Impact Publishers, 2001).

The Four Agreements by Don Miguel Ruiz (Amber-Allen, 1997).

Healing Without Freud or Prozac: Natural Approaches to Curing Stress, Anxiety and Depression by David Servan-Schreiber (Rodale, 2005, 2011).

Mindful Recovery: A Spiritual Path to Healing from Addiction by Thomas Bien and Beverly Bien (Wiley, 2002).

The Power of Now: A Guide to Spiritual Enlightenment by Eckhart Tolle (New World Library, 2003).

Self-Coaching: How to Heal Anxiety and Depression by
  Joseph Luciani (Wiley, 2001).
The Sleep Book: How to Sleep Well Every Night by Guy
  Meadows (Orion, 2014).

BOOKS TO HELP WITH RELATIONSHIP PROBLEMS
CBT

*Love Is Never Enough: How Couples Can Overcome
  Misunderstandings, Resolve Conflicts, and Solve
  Relationship Problems Through Cognitive Therapy* by
  Aaron T. Beck (HarperCollins, 1989).
The Seven Principles for Making Marriage Work by John
  M. Gottman and Nan Silver (Three Rivers Press,
  2000).

BOOKS TO HELP WITH PHYSICAL PROBLEMS AND PAIN
CBT

*Overcoming Chronic Pain: A Self-Help Guide Using
  Cognitive Behavioural Techniques* by Frances Cole,
  Hazel Howden-Leach, Helen McDonald and
  Catherine Carus (Robinson, 2005).
Coping Successfully with Pain by Neville Shone
  (Sheldon Press, 2002).
Coping with Chronic Fatigue by Trudie Chalder
  (Sheldon Press, 1995).

Alternative approaches

*Full Catastrophe Living: Using the Wisdom of Your
  Body and Mind to Face Stress, Pain, and Illness* by Jon
  Kabat-Zinn (Delta, 1990).

Full Catastrophe Living: How to Cope with Stress, Pain and Illness Using Mindfulness Meditation by Jon Kabat-Zinn (Piatkus, 2001).

HELPFUL WEBSITES AND ELECTRONIC RESOURCES

We have identified several additional sources for both self-help and professional help. If you type 'depression', 'anxiety', 'self-help' or similar terms into a search engine, you'll uncover an endless stream of possible resources. Do beware as the internet not only has valuable, reliable resources but also contains thousands of advertisements and gimmicks. Be especially cautious about official-sounding organizations that heavily promote expensive materials. Don't believe absurd promises of instant cures for any problems, psychological or otherwise. In looking at any site do consider and make enquiries regarding the reliability of the source of the information. Check out whether the authors have recognized professional training and qualifications.

Many web forums host chat rooms for people who have a variety of emotional problems. If you do access them for support, bear in mind that you have no idea who you are talking to when you join a web forum. Others in it may know very little about your particular difficulty, or, even worse, be trying to take advantage of a person in distress. However, if you are aware of these possible problems and exercise caution, you may find these forums supportive and helpful. Do be vigilant, and don't get drawn into unhelpful communication which may be masquerading as support.

We've compiled a list of some legitimate websites that don't sell snake oil but do provide excellent information about a variety of emotional issues. Good quality information is most likely to be found on websites provided by governmental, professional and charitable organizations. **We have focused on the UK but most countries will have their own equivalents.**

PROFESSIONAL ORGANIZATIONS

In addition to providing information to the general public, most of the professional organizations below also issue and enforce the codes of practice for their profession. The codes of practice ensure that employers, colleagues, service users, carers and members of the public know what standards they can expect from that registered profession. They protect the public by requiring high standards of education, conduct and practice of all members of the profession.

The British Psychological Society www.bps.org.uk

The BPS provides information about treatment and facts about a variety of emotional disorders. It also holds a register of qualified psychologists, and information on their areas of specialization.

The British Association for Behavioural and Cognitive Psychotherapies www.babcp.com

The BABCP has information for the public about anxiety and other mental disorders. It has a list of

CBT-qualified therapists within a variety of core professions, including clinical psychology, nursing, medicine, social work and psychotherapy.

Royal College of Psychiatrists www.rcpsych.ac.uk
The Royal College of Psychiatrists provides comprehensive factsheets and information.

The Health and Care Professions Council
www.hpc-uk.org
The HCPC is the regulator for specified professions including practitioner psychologists, set up to protect the public. They hold a register of health professionals who meet the required standards for their training, professional skills, behaviour and conduct.

VOLUNTARY ORGANIZATIONS FOR INFORMATION AND SUPPORT
Anxiety UK www.anxietyuk.org.uk
This is a user-led organization, run by sufferers and exsufferers of anxiety disorders, supported by a high-profile medical advisory panel. They work with people suffering from anxiety disorders, providing information, support and understanding via an extensive range of services, including one-to-one therapy, telephone support and groups.

ChildLine www.childline.org.uk
ChildLine is a counselling service for children and young people. They are available by phone on 0800 1111, or via email from the website. They also offer a

one-to-one chat or you can send them a message – you can post messages to the ChildLine message boards and you can text ChildLine. You can contact ChildLine about anything – no problem is too big or too small. Perhaps you are feeling scared or out of control or just want to talk to someone. Some of the things that you might want to be in contact about are feeling lonely or unloved, worries about the future, problems about school, bullying, drugs, pregnancy, HIV and AIDS, physical and sexual abuse, running away and concerns about parents, brothers, sisters and friends and crimes against you.

Mind www.mind.org.uk
Mind is the national association for mental health and it provides high-quality information and advice, as well as running campaigns to promote and protect good mental health for everyone. They have very help-ful factsheets, and information on activities and groups in your area.

The National Society for Prevention of Cruelty to Children www.nspcc.org.uk
The NSPCC is available to provide advice to adults who are worried about a child. Whether you have a serious concern about a child being badly treated which you wish to report, or simply need advice about a child's welfare, they are available to help. You don't have to say who you are and if you are calling they do not track where you are phoning from. The telephone number is 0808 800 5000.

No More Panic www.nomorepanic.co.uk
This site provides valuable information for sufferers and family members of those suffering from panic, anxiety, phobias and obsessive-compulsive disorders (OCD), as well as information for friends and family. They provide members with support, advice and a chance to meet like-minded people and make friends along the way. They advocate using the website information, message forum and chat room alongside any care a sufferer is currently receiving from their physician or therapist.

OCD-UK www.ocduk.org
This organization is run by suffers for sufferers, and works with and for people with obsessive-compulsive disorder (OCD).

Rethink www.rethink.org
This is a national mental health membership charity, helping people affected by severe mental illness. Their aim is to make a practical and positive difference by providing hope and empowerment through effective services, information and support to all those who need such help. They carry out research which informs national mental health policy and actively campaign for change.

Their website includes free PDF guides, factsheets, DVDs, educational packs and other mental health resources. They support people affected by mental illness in getting back into employment. They also give information, and have factsheets on adults' and children's difficulties. Their resources are available in several languages.

SANE www.sane.org.uk

SANE provides information on all aspects of mental illness including depression and manic-depression. The site also offers practical support for anyone affected by mental ill health.

INFORMATION ABOUT MENTAL HEALTH CONDITIONS
The Mental Health Foundation

www.mentalhealth.org.uk

The Mental Health Foundation brings together evidence-based help and information to influence UK policy and practice, and to support people with mental ill health or with learning disabilities. Their aim is to reduce the suffering caused by mental ill health and to help people lead mentally healthier lives. They help people to survive, recover from and prevent mental health problems, by carrying out research, by developing practical solutions for better mental health services, by campaigning to reduce stigma and discrimination and by promoting better mental health for all. They work across all age ranges and all aspects of mental health to maximize everyone's mental wellbeing.

NHS Choices http://www.nhs.uk

This website is the UK's biggest health website, designed to give all the information you need to make choices about your health. NHS Choices is described as the online 'front door' to the NHS. It contains information about hundreds of different conditions, advice on healthy living and support for those looking after others.

National Institute of Mental Health www.nimh.nih. gov NIMH reports on research concerning a wide variety of mental health issues. They provide a range of educational materials on a wide variety of difficulties, as well as resources for researchers and practitioners in the field.

The National Institute for Clinical Excellence
www.nice.org.uk
This is a government organization which makes information available about causes, prevalence, and treatments of disorders of both children and adults. It provides guidance, sets quality standards and manages a national database to improve people's health and prevent and treat ill health. NICE makes recommendations to the NHS on new and existing medicines, treatments and procedures, and on treating and caring for people with specific diseases and conditions. In addition, NICE makes recommendations to the NHS, local authorities and other organizations in the public, private, voluntary and community sectors on how to improve people's health and prevent illness and disease.

Netdoctor www.netdoctor.co.uk
Netdoctor.co.uk is a medical information and health website which is a collaboration between doctors, health care professionals, information specialists and patients who believe that medical practice should be based on quality-assessed information and, wherever possible, on the basis of the principles of evidence-based medicine. They are an independent group of over 250 of the UK

and Europe's leading doctors and health professionals who write, edit and update the website. They also respond to users' questions regarding general health concerns.

WebMD www.webmd.com
This website has both UK and USA sections. It provides a vast array of information on both physical and mental health issues, including information about psychological treatments, drug therapy and prevention.

TREATMENT RESOURCES TO TAKE YOU A STEP FURTHER
Mindfulness based cognitive therapy www.mbct.co.uk
This website provides additional information about mindfulness based cognitive therapy.

Be Mindful www.bemindful.co.uk
This website will help you to find mindfulness courses and resources in your area.

Get Self Help www.getselfhelp.co.uk
This website offers an absolutely vast selection of simply written, practical guides on CBT self-help, as well as information, resources and a wide selection of tools and techniques. It also includes therapy worksheets so you can track your progress. It's designed and administered by a qualified CBT therapist with years of experience of working with people with mental health problems.

The British Holistic Medical Association www.bhma. org This organization offers free podcasts on a number of topics including managing stress, wellbeing and work-related stress. They also provide information on health problems, mainstream and complementary treatments and self-care. In addition they are developing factsheets on common conditions for which complementary therapies and mind-body medicine may be of benefit. The organization is interested in the supporting evidence base and is open about where this may exist and where it does not. They also offer a range of self-help tapes.

Living Life to the Full www.llttf.com
This website provides an interactive self-help life skills training package based on CBT for those with mild to moderate depression and anxiety. It also has a number of free downloadable resources.

The Mood Gym www.moodgym.anu.edu.au
This is another free online self-help programme based on CBT. Again, this is likely to be most suitable for people with mild to moderate mental health difficulties. This is an Australian website but the online help is relevant to people in other countries.

Beating the Blues www.beatingtheblues.co.uk
This is an online CBT treatment package for depression, recommended by the NHS for people with mild to moderate symptoms. It has a solid and established evidence base. This is not free to individual patients but can be accessed free of charge through IAPT and other psychological therapy services throughout the UK (see

the 'Finding professional help' section earlier in this chapter). Ask your GP about this.

Fear Fighter www.fearfighter.cbtprogram.com
Like Beating the Blues this is an online CBT package with a strong evidence base recommended by the NHS. Again, it has to be accessed through NHS services such as IAPT and is less regularly offered than Beating the Blues. Discuss with your GP which services might offer this.

ONLINE VIDEOS ON CBT AND MINDFULNESS
The following are online videos which are used by a number of CBT therapists when teaching people about the advantages of using mindfulness (see Chapters 4 and 6) with more traditional CBT approaches. Just enter the whole address into your web browser. Many of the above resources will also provide multimedia content.

PHILIPPE GOLDIN – 'COGNITIVE NEUROSCIENCE OF MINDFULNESS MEDITATION'
https://youtu.be/sf6Q0G1iHBI
This video shows how attending to the here and now can reduce distress. It gets into the more technical aspects of mindfulness and explains the brain and mind functions that meditation (attending to the present) works on. He describes the **narrative self** – and how we can believe the stories that we make about ourselves more than actual reality. Though it's a bit technical at times, it's worth watching all the way through.

JON KABAT-ZINN – 'COMING TO OUR SENSES'
https://youtu.be/qvXFxi2ZXT0
This video explains how people lose touch with themselves and get wrapped up in destructive behaviours and thoughts. If you can get through the lengthy introduction (and the poetry – not everyone's taste) the discussion itself can be inspiring and yet stays really simple.

Index

A PRACTICAL GUIDE TO ASSERTIVENESS

What is assertiveness, and what are benefits? Filled with straightforward, practical advice, *Introducing Assertiveness: A Practical Guide* will help you find out, allowing you to overcome passive behaviour and take ownership of your own thoughts and feelings without becoming aggressive.

Experienced life and business coach David Bonham-Carter provides clear, practical steps to help you develop they key characteristics of assertiveness – steps that can improve your work life and your personal life.

ISBN: 9781785783319
£7.99

A PRACTICAL GUIDE TO HAPPINESS

Apply the wisdom of philosophers to become a happier person.

What is happiness? What makes you happy? Is there more to life than happiness?

Learn to cultivate your taste for pleasure, free yourself from the various disturbances of life, and overcome irrational expectations that cause distress. Go with the flow and rediscover the joy of existence.

Filled with exercises, tips and case studies, this Practical Guide will enable you to see happiness in a new light, with the help of the world's greatest minds

ISBN: 9781785783241
£7.99

A PRACTICAL GUIDE TO POSITIVE PSYCHOLOGY

Start looking on the bright side.

With expert encouragement and guidance, you will set out your own positive psychology experiment to discover your strengths, overcome negative attitudes, focus on what gives you purpose, and take control of your life choices.

From savouring positive emotions to building better relationships and developing resilience, you will gain the tools to boost your mental and physical well-being and to find fulfilment in everyday life. This is the perfect concise start to making your life better.

ISBN: 9781785783852
£7.99

A PRACTICAL GUIDE TO SPORTS PSYCHOLOGY

Improve your performance and achieve your goals.

Become fitter and stronger, relieve stress and anxiety, and recover from injury faster and more effectively with tips from sport psychology.

Whether you're looking to make exercise a habit or to perform better in your sport – whatever your level –this Practical Guide employs proven techniques to train your mind, help you to feel better and smash your exercise goals.

ISBN: 9781785783272
£7.99

A PRACTICAL GUIDE TO NLP

Remap your path to success with effective motivational techniques.

Learn to use the psychological techniques of neurolinguistic programming (NLP) to 'reprogram' your brain – replacing the negative attitudes that hold you back with positive thought patterns that will enable you to be more effective, confident and successful.

In just under twenty simple steps, Neil Shah shows you how to use NLP to develop new habits of behaviour and thought that will help you succeed in all areas of life, from influencing others and understanding how they influence you, to achieving your goals, to managing stress.

ISBN: 9781785783906
£7.99